Acknowledgements

This report was prepared as part of a research project entitled **Review of Planning Policy for the Coast and Earth Science Information in Support of Coastal Planning and Management**. The project was funded by the Department of the Environment under its planning and research programme (contract No. PECD 7/1/403).

The results of the first part of this study were published in 1993 as:

Rendel Geotechnics, 1993. Coastal Planning and Management: A Review HMSO.

This report addresses the second part of the study and is concerned with the use of the earth science information in the coastal planning and management. The report was written by Mr. E.M. Lee of Rendel Geotechnics with guidance provided by Dr A.R. Clark (Rendel Geotechnics) and Dr J.C. Doornkamp (Nottingham University).

Background research and specialist advice has been provided by:

Mr. D. English	Rendel Geotechnics
Dr. R. Moore	Rendel Geotechnics
Mr. P. Phipps	Rendel Geotechnics
Mr. A.S. Freeman	Rendel Planning and Environment
Prof. D. Brunsden	King's College, University of London
Mr T.M. Dibb	Consultant GIS Specialist
Dr. P. Doody	Consultant
Dr. R Haines Young	Nottingham University
Prof. J. Pethick	Institute of Estuarine and Coastal Studies, University of Hull.
Dr G. Priestnall	Nottingham University

Department of the Environment

Coastal Planning and Management: A Review of Earth Science Information Needs

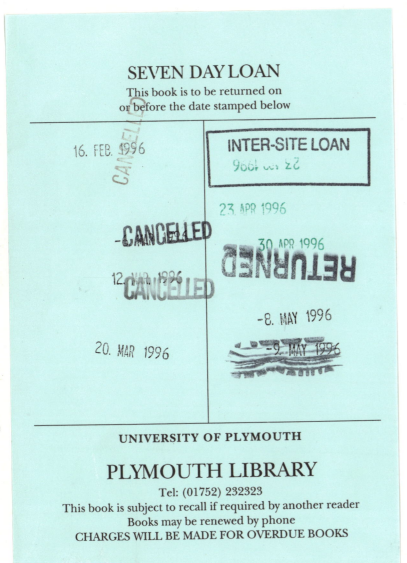

3

HMSO: LONDON

This publication has been produced from
Camera Ready Copy supplied by the Department of the Environment

Executive Summary

The requirement for earth science information

The coastline is an important part of our natural heritage. It provides outstanding scenery, plant and wildlife habitats, and geographical features. The coast also provides significant benefits of shelter and deep water for ports and harbours, breeding grounds for fish and shellfish, and opportunities for tourism and recreation. The developed coast has its share of dereliction and of land which will need to be re-developed in the foreseeable future. The sea-bed is an important source of sand and gravel for the construction industry and for coastal defence schemes.

The coastline is a constantly changing environment which is subject to the rise and fall of the tides, movements of currents in the sea, and storms. These processes give rise to natural hazards which may affect development undertaken in vulnerable locations. Conversely, development may affect the natural processes, leading to unforseen consequences elsewhere, or may conflict with conservation interests.

In many instances planning and development decisions are influenced by social, economic and environmental factors. However, there are cases where the nature of the ground and the operation of physical processes will be relevant to the safe, cost-effective development of the land. This is especially so on the coast where the dynamic nature of the environment dictates that planners and developers need to be aware of physical factors such as erosion, deposition and flooding, mineral resources and conservation features.

Although planning considerations will not be uniform aground the country, it is possible to identify a number of key issues that will need to be addressed on most coasts. These include:

(i) **Development**

- the impact of erosion, deposition and flooding processes on development and the need for remedial and defence works;

- the effects of development on the operation of physical processes along the coast;

- the effects of development on mineral, water and conservation resources.

(ii) **Conservation**

- the importance of physical processes in creating and maintaining conservation features;

- the effects of conservation policies on diverting development into vulnerable areas away from conservation resources.

(iii) **Recreation**

- the importance of physical processes in creating and maintaining recreation resources such as beaches and sand dunes;

- the effects of recreational facilities on the stability of features such as sand dunes.

(iv) **Minerals**

- the occurrence and significance of mineral resources such as aggregates for construction and coastal defence, and building stone.

- the effects of mineral extraction on the operation of physical processes along the coast.

- the effects of mineral extraction on marine and coastal resources;

- the conservation opportunities provided by mineral workings.

These are not simply local concerns, with the influence of physical processes often extending well beyond a single authority area.

Earth science information could be used more often and to a greater effect by coastal managers and planners in order to ensure that development:

- is not located in areas which are at risk from erosion, land instability, flooding or the deposition of sediment;

- does not affect the natural balance of the coastline to the extent that erosion is caused elsewhere or that costly coastal defences have to be constructed and maintained; and

- is undertaken in a sustainable manner with due regard to the environment.

It is clear, in particular, that the construction of coast defence works has often led to a false sense of security that land has been fully protected against a hazard whereas, in reality, the hazard has only been reduced. In such circumstances, planning permission has sometimes been given to development thus creating potentially greater losses if the defence is overcome.

In order to consider and resolve such issues there is a need for good information on geology, soils, surface and groundwater, and shoreline and near-shoreline processes. Three key aspects need to be addressed:

- risks
- sediment budget
- sensitivity

Risks to new and to existing development and to specific uses of land need to be identified. Consideration is needed of the extent to which these may be overcome and examination of levels of risks are required to identify priorities for action.

The **sediment budget** needs to be examined because of the significance of the natural movement of sediment around the coast in maintaining natural coastal defences such as beaches, sand dunes, shingle ridges, mudflats and saltmarshes. These movements are also crucial to the maintenance of conservation sites and recreation areas. It is important, therefore, to examine the potential effects of any interference on the operation of processes on neighbouring coasts.

The **sensitivity** of features and of coastlines to changes is variable whether the changes are due to human activity, short term natural processes, or long term factors such as global warming and sea level rise.

Whilst all of these factors have some bearing on management and planning on the coastal zone, the detail of information required depends on the level within the planning and development process. Much of the relevant information is being compiled already as part of the basis for shoreline management plans. It is also important to harmonise policies and provisions within development plans with those in management plans.

Approaches to Investigation

The type and quantity of information needed to support decisions by planners and developers will vary according to the stage in the decision-making process. For both planners and developers the requirement for information ranges from general awareness of the character of an area to site specific information. As a result a range of approaches are relevant of different stages in the planning and development process (Figure 1).

(i) **General Assessment**; for many purposes, such as strategic policy formulation and preparing local plan proposals, a broad brush approach can contribute significantly to the safe, cost-effective development and use of land. General assessment of the principal physical conditions in an area can provide a relatively quick appraisal through the collection and interpretation of readily available data sources. General assessment should be undertaken by the planning authority as part of its survey of the principal physical characteristics of its area when preparing or reviewing a development plan;

(ii) **Site Reconnaissance**; both planners and developers have an interest in the selection of sites for development and thus may need to undertake appropriate site studies;

Figure 1 The sequence of investigations in the planning and development process.

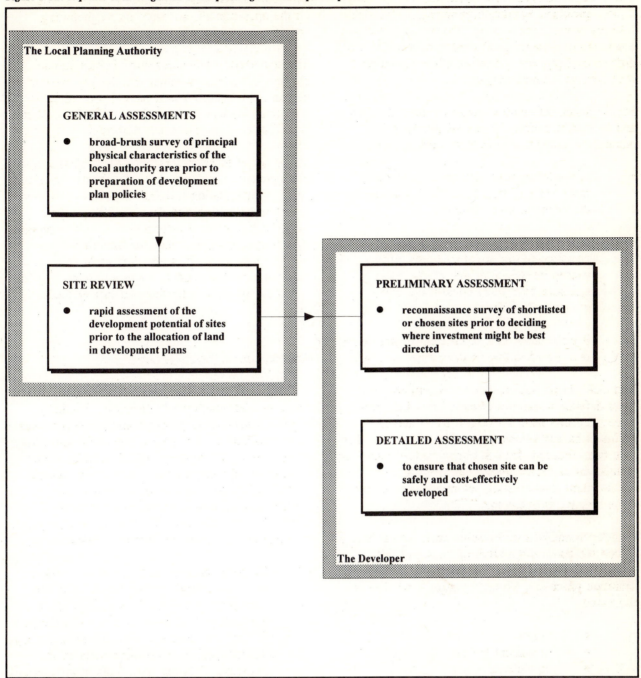

A **site review** is necessary to provide **planners** with as quick appraisal of potential sites for development and redevelopment, ensuring that, for example, they can be safely developed. Indeed, the nature of the "plan led" planning system dictates that development plan proposals should take full account of potential problems as failure to do so could lead to development in unsuitable locations;

This level of investigation may also provide advance warning to **developers** of the potential difficulties associated with particular sites and an indication of the cost implications (i.e. **preliminary assessment**);

(iii) **Detailed Assessment**; to demonstrate to the local planning authority that the proposed development is suitable and takes full account of all material planing considerations, including, as appropriate, physical, economic and environmental factors. This level of investigation should enable developers to assess the level of risks associated with the development and the likely precautionary measures such as flood defence.

It is important to ensure that the effort spent by planners is appropriate to the level of investigation. In many instances it is not feasible to collect and collate all existing data relevant to any study, partly because some holders will not provide it, partly because considerable time may be needed to access widely dispersed data sets. Limitations to the time and resources available to planners will generally dictate that comprehensive geographical of thematic coverage is unlikely to be a realistic option. Investigations will, therefore, be a practical compromise addressing **key issues** that relate to specific policy needs, with effort directed towards areas where there is significant pressure for development.

The collection and use of earth science information is to support policy needs and site specific decisions. These will vary with the land use character of the coastline and the extent to which there is significant pressure for development or redevelopment. From this perspective the coast may be divided broadly in to four land use types:

- the undeveloped coast, conserved both for its landscape value and for its nature conservation interest;
- other areas of undeveloped or partly developed coast;
- the developed coast, usually urbanised but also containing other major developments (e.g. ports, power stations, etc.); and
- the despoiled coast, damaged by dereliction caused by mining, waste tipping and former industrial uses.

The interactions between the coastal land use and the type of environment are critical in defining the amount of information required for decision making. The amount of information required to support planning on much of the undeveloped coast will generally be less than on similar but developed sections where, for example, more effort needs to be directed towards considering the degree of risk. Site reconnaissance studies and detailed assessment will be required most frequently in considering allocations of land for development on the **developed** or **despoiled** coast. General assessments, on the other hand, will need to be undertaken for the whole of the coastline of an administrative area, thereby providing a comprehensive planning base for the whole area so that the full range of options for development and conservation can be considered.

Strategic planning

When revising or replacing development plans the local planning authority should (Figure 2);

- undertake a general assessment of physical conditions along the coastline, through the collection and interpretation of readily available data, as part of the survey of the principal characteristics of their area. This should identify key environments such as estuaries, coastal lowlands, coastal cliffs, and sand dunes;

- identify the key issues which have implications for planning in each environment, for instance flooding, channel erosion, sedimentation, landslides, cliff recession and wind erosion) and to establish hazards, risks, and sediment budgets, as appropriate, for each area;

- assess the environmental resources of the area, such as the role of erosion, deposition and flooding in maintaining wildlife habitats and in reducing rates of erosion elsewhere;
- to determine which of the factors identified are of great enough significance to require consideration in local plans and Part II of Unitary Development Plans (UDPs).

In a Structure Plan or a UDP Part I it will normally be sufficient to identify the factors which may need to be taken into account in more detailed strategic planning. Greater detail is required for surveys of principal characteristics of areas covered by local plans or a UDP Part II.

The information needs in different coastal environments are summarised in Chapters 6-9. Much of the information required may be provided by the current programme of shoreline management plan preparation although the information may need to be edited, adapted and re-organised for planning purposes. Coastal groups and organisations such as the NRA should be consulted in the course of the general assessment of the area and on the draft polices arising from it.

In some cases it may be appropriate to use a computerised database for handling information, and to consider whether a geographic information system offers opportunities for information handling and display. If such facilities are developed there is sound sense in making these

Figure 2 Forward planning: the need for earth science information.

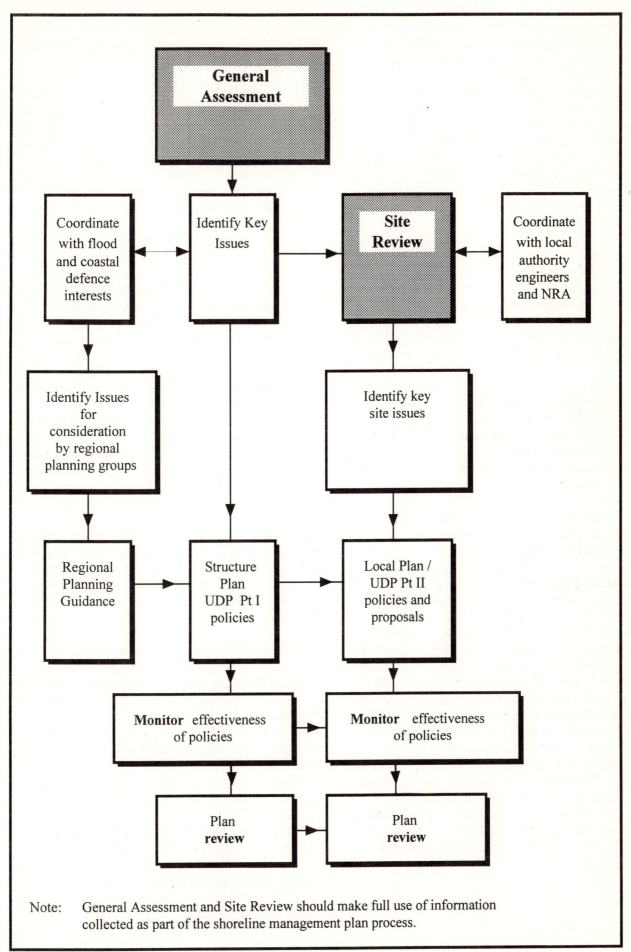

Note: General Assessment and Site Review should make full use of information collected as part of the shoreline management plan process.

publicly available so that they can be consulted developers and by planners in respect of the development control process.

Land Allocations

Where allocations of land are identified by local planning authorities in Local Plans and UDP Part II it is important that they in should, general terms, be suitable to be safely developed and the issues which will need to be taken into account by prospective developers should be identified. Therefore local planning authorities should carry out a site review prior to the next review of the development plan in order to determine for land which may be allocated for specific uses:

- the nature of the ground and any processes which may affect it;

- the significance of any constraints or opportunities for planning and use of the site and, in particular whether development of the site might adversely affect environmental resources or might lead to increased coastal erosion and, thus, a greater need for coastal defences, elsewhere; and

- the suitability of the land for development and the nature of any coastal defences which might be required before development may proceed.

Information and advice should be sought from local authority coastal engineers and the NRA when assessing the suitability of any coastal land for specific types of development (as described in Chapters 6–9). Liaison with these bodies should also consider:

- whether any prospective land allocations might conflict with provisions of shoreline management plans; and

- the level of detail of reports which should be submitted by developers in support of any planning applications.

The results of the site review should be made available to prospective developers in order to assist in site selection and planning of site investigations,

Selection of Sites

Prospective developers should have regard to the provisions of the local plan when selecting possible development sites. Any specific issues identified in the plan should be noted and steps should be taken to weigh these up before deciding on a specific site. In particular, it should be noted that identified issues will need to be thoroughly explored in any site report or environmental assessment.

It is prudent to discuss development proposals at the earliest possible stage with the local planning authority in order to determine whether potential problems may exist and what might be done to overcome these. In particular, it is important to determine those circumstances in which it may be necessary to undertake a site investigation before a planning application is submitted. Developers and their advisors should also have regard to any shoreline management plan.

Planning Applications

It is the developers's responsibility to determine that a site is suitable for the purpose to which it is proposed that it should be put. It is in the developer's interests, therefore, to determine whether the site is in an area which may render it subject to problems which may affect the value of the land and of any development upon it.

In areas defined by the local planning authority as being subject to potential problems, the developer should prepare a site report taking account of the perceived problems, unless there is good evidence to the contrary, and submit it in support of any planning application. In some cases a more extensive environmental assessment may be required (Figure 3).

Any report should be prepared by a suitably qualified expert, or experts, and should take account of all of the relevant physical, environmental and economic factors. Particular issues which may be relevant include whether:

- the land is capable of supporting the load to be imposed;
- the site may be threatened by erosion, land instability or flooding;
- the development might affect the level of risk in adjacent areas, or elsewhere along the coast;

Figure 3 The role of the developer in the investigation of sites.

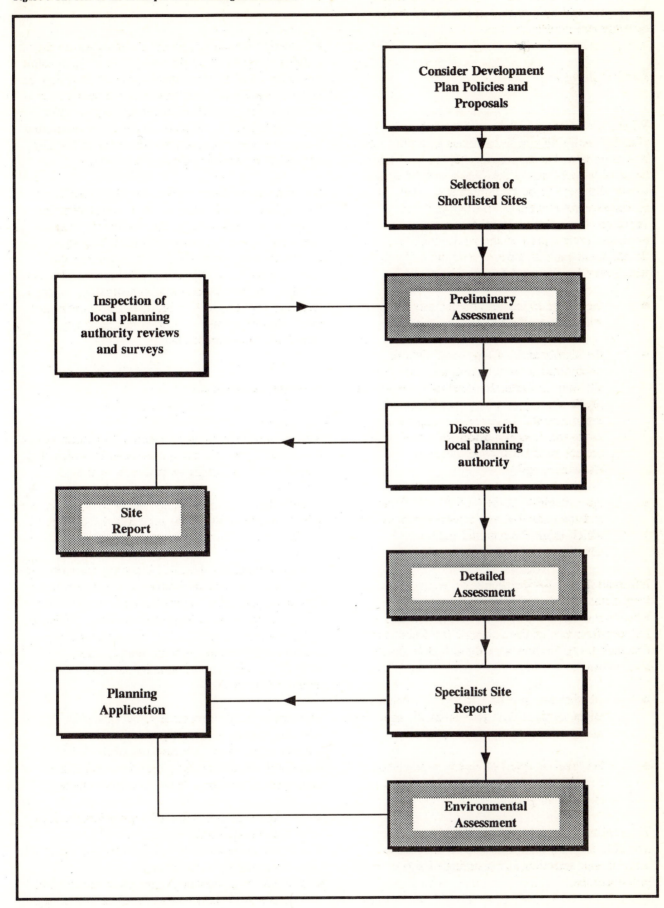

- the nature of the coastal defence scheme options which might overcome potential problems or any potential impacts;
- the mitigation measures that might be used to reduce any undesirable aspects of the proposals.

Government policy is to avoid placing an additional burden of responsibility on future generations by increasing unnecessarily the numbers of areas which need artificial protection against erosion or flooding. For that reason, any requirements for new, or improved, coastal defences at future development sites should be provided at the developer's expense.

The local planning authority will, where appropriate, need to (Figure 4):

- ensure that the developer submits a site report, prepared by appropriate experts, which addresses all of the key issues at a level of detail appropriate to the potential severity of any problems and which is adequate for the determination of the application;

- request additional information if the presented report is inadequate in order to have an adequate basis for determining the application;

- seek advice from coastal engineers and the NRA about a the implications of the results of site reports, especially with regard to possible planning conditions or modifications to proposed designs; and

- maintain a list of professional organisations which can be approached by developers when trying to identify suitable experts for carrying out necessary site evaluations.

Co-operation

Liaison and sharing of information between local planning authorities, coastal defence groups, the NRA, coast protection authorities, and conservation interests is important in order to develop a coordinated approach to management of coastal issues. It is also in the interests of the economy and efficiency that local planning authorities should make full use of the information which is being compiled as part of the preparation of shoreline management plans. Similarly, developers should discuss development proposals with the local planning authority at the earliest possible opportunity and should be able to make use of databases developed by local authorities for coastal planning.

However, many coastal processes operate over a broader scale than individual local authority boundaries thus sustainable development can be achieved only through an awareness of the behaviour of the large sediment transport systems which operate around the coast. In this context, regional planning groups should identify those issues which need to be considered over a wider area than the individual authority area. Important examples include the effects of development on the natural movement of coastal sediment and the potential impacts of development on natural coastal defences and on conservation interests elsewhere.

Trial Study Areas

Five trial studies have been carried out to demonstrate how earth science information can be collected and collated as part of a review or survey of a planning area. The areas have been selected to demonstrate the types of information that can be readily accessed to support the preparation of development plan polices in a range of coastal environments, i.e. the studies are the equivalent to a general assessment. The areas chosen are:

- Fraserburgh – Scotstown Head (Grampian)
- Seaham – Teesmouth (Durham and leveland)
- Filey – Scarborough (North Yorkshire)
- Flamborough Head – the Outer Humber Estuary (Humberside)
- Cromer – Sea Palling (Norfolk)

The results of the studies are held on open file at the Department of the Environment and can be accessed on request. Each one is of a limited length of coast, although it is hoped that the methods demonstrated could be practised over whole administrative areas in advance of development plan reviews. The approach is based on **Applied Earth Science Mapping** techniques that have been developed and promoted by the Department of the Environment over the last decade or so. The studies draw upon planners' familiarity with maps at a general scale and involve the production of a combination of thematic maps

Figure 4 Control of development: the need for earth science information.

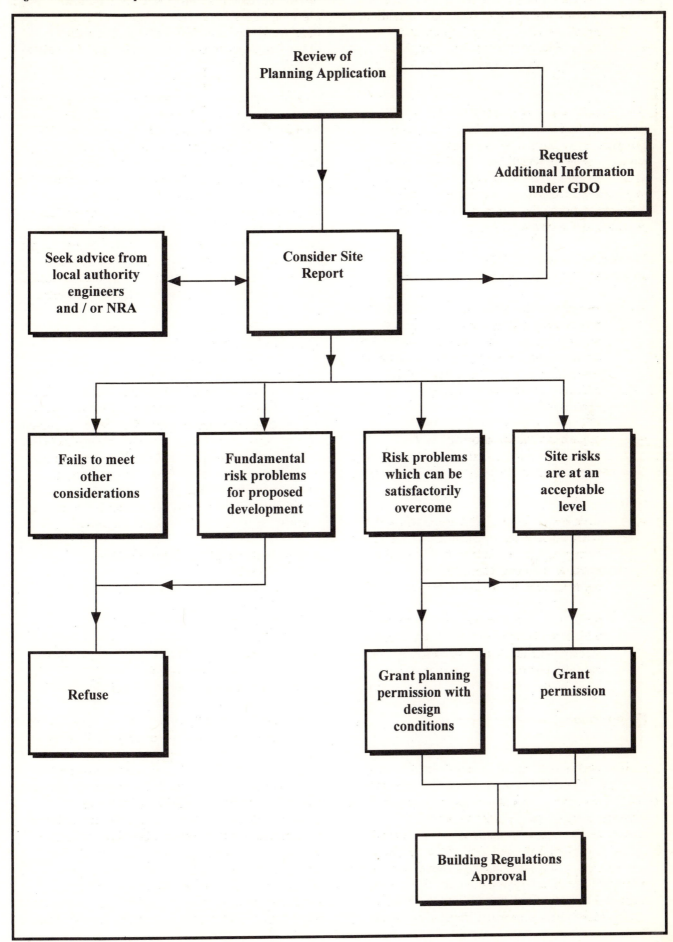

which become increasingly focused on key planning issues as they are developed from the basic factual information.

The studies have all involved:

- identification of **key considerations** within different environments occurring along the stretch of coast;

- identification of **key information sources** associated with different considerations;

- compilation of an **inventory** map of coastal geomorphology;

- preparation of **interpretative** maps of either risks, sediment budget or sensitivity;

- compilation of **summary** maps identifying resources and constraints;

- preparation of preliminary planning guidance;

- preparation of a short non-technical report to support the maps, including more detailed technical information in Appendices.

Information Management

Effective management of the earth science information collected and compiled by local authorities and other organisations should be viewed as essential for ensuring its efficient use by both planners and developers. Traditional forms of information management have involved indexed archives of reports and maps (**manual systems**). Although such approaches are unlikely to become redundant, it is becoming apparent that **computer-based systems** have considerable potential for enhancing:

- the storage and retrieval of information;
- the accessibility of information to decision-makers;
- the updating of reports, interpretative maps and revision of planning and guidance.

There have been considerable advances in computer technology, although it should not be assumed that this will always be the most cost-effective route for local authorities. Indeed, the key to successful information system development lies in gearing the development to;

- the perceived users of the system (in terms of background and expertise);

- the information and knowledge that those users wish to extract from the system;

- the time constraints of development.

Appropriate information systems which fulfil a particular set of functions can be located at a point along a spectrum which represents the range of possible information systems. Systems developed within this spectrum can be generally classified into one of the following classes:

- totally manual systems;
- simple databases;
- geo-referenced databases;
- complex geographic information systems.

Chapter 10 examines the development and use of computerised databases, highlighting the relative attributes of a range of systems of different levels of sophistication. The role of Geographic Information Systems (GIS), the most sophisticated of available systems, is examined with reference to a demonstration system as part of this study (the Ventnor GIS).

Need for Guidance

It is recommended that the Department of the Environment should draw the attention of planning authorities and developers to the results of the study in respect of underpinning part of the implementation of PPG20, and should take account of the findings in any future revision of that guidance. In addition, it is felt that there could be a role for Regional Planning Guidance issued by the Department in highlighting wide ranging issues which may need to be taken into account by planners and managers in the coastal zone.

Figure 2.2 Summary of issues: resource management (after Rendel Geotechnics, 1993).

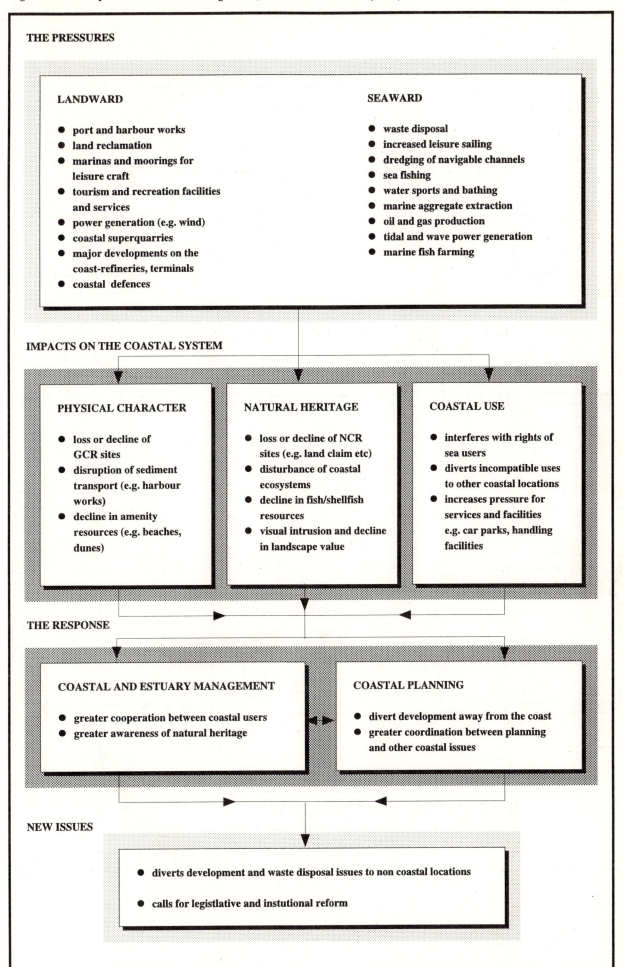

THE PRESSURES

LANDWARD

- port and harbour works
- land reclamation
- marinas and moorings for leisure craft
- tourism and recreation facilities and services
- power generation (e.g. wind)
- coastal superquarries
- major developments on the coast-refineries, terminals
- coastal defences

SEAWARD

- waste disposal
- increased leisure sailing
- dredging of navigable channels
- sea fishing
- water sports and bathing
- marine aggregate extraction
- oil and gas production
- tidal and wave power generation
- marine fish farming

IMPACTS ON THE COASTAL SYSTEM

PHYSICAL CHARACTER

- loss or decline of GCR sites
- disruption of sediment transport (e.g. harbour works)
- decline in amenity resources (e.g. beaches, dunes)

NATURAL HERITAGE

- loss or decline of NCR sites (e.g. land claim etc)
- disturbance of coastal ecosystems
- decline in fish/shellfish resources
- visual intrusion and decline in landscape value

COASTAL USE

- interferes with rights of sea users
- diverts incompatible uses to other coastal locations
- increases pressure for services and facilities e.g. car parks, handling facilities

THE RESPONSE

COASTAL AND ESTUARY MANAGEMENT

- greater cooperation between coastal users
- greater awareness of natural heritage

COASTAL PLANNING

- divert development away from the coast
- greater coordination between planning and other coastal issues

NEW ISSUES

- diverts development and waste disposal issues to non coastal locations
- calls for legistlative and instutional reform

13

"coastal zone management delivered through a cascade of national, regional and local Coastal Zone Plans is the key to sustaining the present uses, enjoyment and ecological richness of the coastal zone into the future"

(House of Commons Environment Committee, 1992)

Whilst supporting the preparation of management plans, the Government has, however, advocated an approach that concentrates on resolving specific local conflicts on stretches of coast where management pressures exist. Management plans have, therefore, tended to be focused on a particular range of themes reflecting the responsibilities of the lead agencies. A guide to coastal planning and management initiatives in England has been prepared by the National Coasts and Estuaries Advisory Group (NCEAG, 1994). This guide reveals an immense diversity of plans, reflecting local complexities and the wide range of interests and responsibilities (Figure 2.3).

Amongst the most important types of plan are:

(i) **Shoreline management plans;** prepared by operating authorities to address strategic coastal defence issues over specified lengths of coast such as a littoral sediment cell or sub-cell (Figure 1.3). In England and Wales these plans are generally prepared through coastal defence groups involving coast protection authorities and the NRA (Figure 1.4). Interim guidance on plan preparation was issued in 1994 (MAFF, 1994) and it is anticipated that coastal defence groups will begin producing plans as soon as practicable;

(ii) **Coastal management plans;** prepared by local authorities, conservation agencies and other interest groups for fairly local stretches of coast, often embracing land and water-based issues. The **South Wight Coastal Zone Management Plan**, for example, addresses a wide range of issues from landscape, conservation, tourism and recreation to coastal defence (South Wight BC, 1994);

(iii) **Heritage Coast and AONB Coast management plans;** prepared by local authorities and supported by the Countryside Commission and its equivalents in Wales (CCW) and Scotland (SNH). These plans usually exhibit a strong conservation emphasis;

(iv) **Estuary and Harbour management plans;** often prepared by groups of local authorities in associated with bodies such as harbour authorities or conservancies. They frequently address resource management and water use issues e.g. the **Chichester Harbour Amenity Area Management Plan;**

(v) **Catchment management plans;** prepared by the NRA in England and Wales to examine all the water-related management aspects of individual river catchments, including estuaries and, occasionally, adjacent coastal waters (eg. the **River Esk and Coastal Catchment Management Plan**; NRA, 1994).

(vi) **Water level management plans;** prepared by drainage authorities in England and Wales to establish an agreed balance between agriculture, flood defence and conservation in wetland areas. Conservation guidelines issued by MAFF state that plans should be produced for areas where water levels are managed, with priority given to SSSIs (MAFF/DoE/Welsh Office, 1991; see also MAFF et al, 1994).

The Nature of the Planning System

The planning system aims to regulate the development and use of land in the public interest. Planning powers are exercised by local planning authorities whose most important functions are:

● the preparation of development plans;
● the control of development, through the determination of planning applications.

Figure 2.3 Coastal management initiatives in England (after NCEAG, 1994).

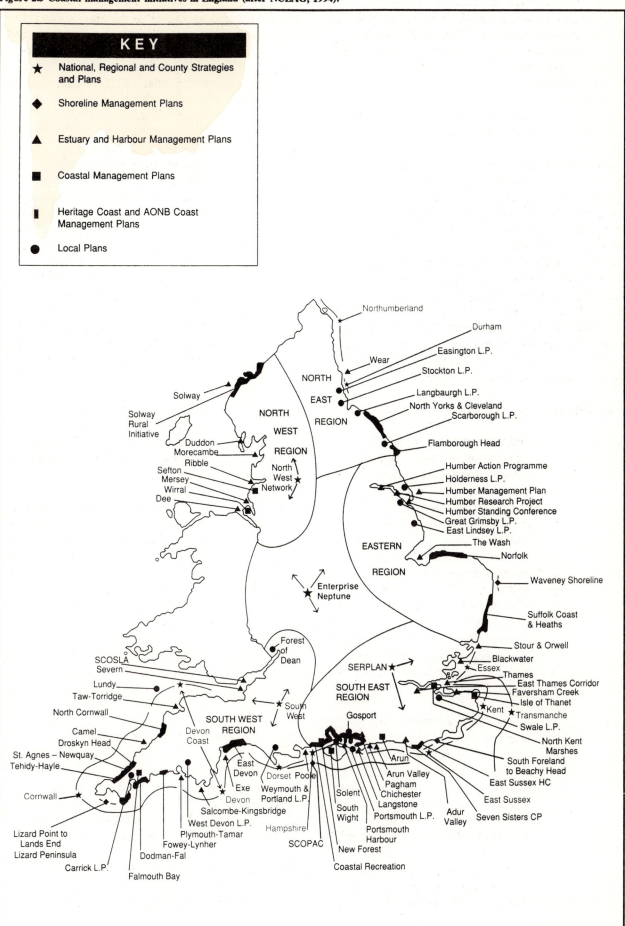

The planning system can be described as "plan-led" in that all planning decisions (either by local authorities or the Secretary of State) must be made in accordance with the development plan, unless material considerations indicate otherwise. This has, in effect, introduced a presumption in favour of those development proposals which conform with the development plan; policy omissions, therefore, can result in development proceeding in areas which might not be suitable.

In both the preparation of plans and handling planning applications, local authorities are required to **consult** with a wide range of interested bodies (e.g. the NRA and conservation agencies and are advised to take account the Governments policy guidance. Of direct relevance to the coastal zone include:

- PPG 20 Coastal Planning (England and Wales; DoE, 1992a)
- North Sea Oil and Gas: Coastal Planning Guidelines (Scotland; SDD, 1974)
- National Planning Guidelines: Priorities for Development Planning (Scotland; SDD, 1981).

In determining planning applications, local planning authorities must take into account representations made in response to statutory consultation or publicity. Permission may be granted subject to such conditions as the local planning authority or the Secretary of State may think fit, provided the conditions are necessary, relevant to planning, relevant to the development, enforceable, precise and reasonable in all other respects. Where matters which are necessary in planning terms cannot be dealt with by way of conditions, such as where the action is not reasonably within the power of the applicant to secure, legal agreements under the planning legislation may be a necessary precursor to the granting of planning permission.

Applicants may appeal to the respective Secretary of State against a decision to refuse permission or grant it subject to conditions. The Secretaries of State may require applications to be referred to them for decision; this **call-in** power is not frequently exercised and is generally only used where planning issues of more than local importance are involved.

The statutory planning system is one of the main instruments for taking account of the EC Directive on Environmental Assessment (Directive 85/337/EEC) in the decision making process, under the **Town and Country Planning (Assessment of Environmental Effects) Regulations 1988.** The regulations apply to certain projects that require planning permission under the **Town and Country Planning Act 1990.** The relevant projects are listed in 2 Schedules corresponding to the Annexes from the EC directive. For Schedule 1 projects environmental assessment (EA) is mandatory; for Schedule 2, EA is required if there are likely to be significant environmental effects (see Table 2.1 for selected examples of projects relevant to the coastal zone). DoE Circular 15/88 (WO 23/88) provides guidance to local planning authorities for assessing the significance for Schedule 2 projects that require planning permission:

- whether the project is of more than local importance;
- whether the project is intended for a particularly sensitive or vulnerable location;
- whether the project is thought likely to give rise to particularly complex or adverse effects, for example, in terms of the discharge of pollutants.

However, these are only general guidelines. In addition, DoE Circular 15/88 sets out indicative criteria and thresholds for some Schedule 2 projects. It is the responsibility of the **competent authority** to make a judgement on the need for an EA on the basis of the indicative criteria and other relevant factors such as the nature, scale and location of a project. Similar procedures apply in Scotland (see **SDD Circular 13/1988 Environmental Assessment Implementation of the EC Directive).**

Not all development requires specific planning permission. In England and Wales, the **Town and Country Planning General Development Order 1988** (GDO; as amended) gives general planning permission in advance for 28 categories of defined classes of development set out in Schedule 2 to the Order. Such permitted rights can be withdrawn by a direction under Article 4 of the GDO which normally requires the Secretary of States approval. However, it is intended that these rights should only be withdrawn in exceptional circumstances.

The Role of the Planning System in Coastal Management

The planning system can be used in a variety of ways to achieve coastal management objectives, including:

Table 2.1 Selected examples of projects covered by the Town and Country Planning (Assessment of Environmental Effects) Regulations 1988.

Schedule 1: EA required in every case.

- crude oil refinery
- thermal power station (over 300 megawatts)
- chemical installations
- ports
- waste disposal installations

Schedule 2: EA required if projects are likely to have significant environmental effects.

- sand and gravel extraction
- thermal power station (less than 300 megawatts)
- shipyards
- chemical installations
- industrial estates
- urban development projects
- harbours
- flood relief works
- coast protection
- oil and gas pipeline installations
- yacht marina
- waste water treatment plant
- a site for depositing sludge

- avoiding locating new development in areas at risk from coastal processes or ensuring that adequate precautions are taken;
- ensuring that development does not affect the level of risk in neighbouring areas or lead to adverse effects on environmental interests;
- supporting coastal management initiatives through the inclusion of appropriate land use policies within development plans.

The system has been effective in arresting the spread of piecemeal development along the coast and providing protection for nature conservation sites above Low Water Mark, through inclusion of land use policies within development plans. It is clear, however, that the role of the system in risk management has not been fully exploited. Tables 2.2 and 2.3 summarise the local planning authority responses to risks in areas of coastal erosion and flooding, respectively (as of 1994, although the situation is rapidly changing). Many authorities have no planning strategy whereas others have developed sophisticated responses in association with the coastal defence interests. The reasons why the system has been less effective in addressing risk issues are likely to be complex, reflecting a combination of planning guidance available at the time of plan preparation and local planning

authority perception of the issues. Here, it is important to note that in the absence of guidance to the contrary many local planning authorities took the view that risks were technical matters for consideration by the developer. Indeed, the situation prior to the mid 1980's can be summarised by:

> "it is generally held that economic considerations e.g. the feasibility of the proposed development, is a matter for the developer, not for the planning authority.
>
> In this connection it could be considered that the extra costs which should be incurred in site investigation, land stability and protection works are not land use planning matters",
>
> (J.S.Turner, Planning Appeals Commission, 1987)

In recent years, however, the Government has emphasised the need for these problems to be taken into account in development plans in England and Wales, through:

- PPG 14 Development on Unstable Land (DoE, 1990);
- PPG 20 Coastal Planning (DoE, 1992a);
- DoE Circular 30/92 (MAFF FD 1/92; WO 68/92) Development and Flood Risk (DoE, 1992b).

Draft advice on flood risk in Scotland has recently been issued (Scottish Office, 1995).

The past reluctance of some local planning authorities to consider risks a planning issue reflects the concern that many damaging events are difficult to predict and that hazards maps need to be prepared with great care to avoid problems such as adverse effects on property values or litigation. However, any moves towards greater precaution and tighter development control to ensure risks are fully considered in the planning process needs to be based on a sound knowledge of the coastal environment, i.e. planning needs to be supported by adequate earth science information.

Table 2.2 Local planning authority responses to unstable coastal cliffs in Great Britain (as of 1993).

AREA OF COASTAL INSTABILITY	GENERAL DESCRIPTION	COUNTIES	STRATEGIC POLICIES	COMMENT, INCLUDING SELECTED DISTRICT LOCAL PLAN POLICIES
Suffolk Coast	Rapid erosion of 10m high sandy cliffs from Dunwich to Southwold: Dunwich now consists of 30–40 houses, having been a small city around 1000 years ago. Cliff erosion problems at Easton Bavents near Southwold.	Suffolk	Yes	Structure plan policy: • development not acceptable which would be likely to be affected by marine erosion in its lifetime. **Suffolk Coastal DC;** local plan policies: • stability will be a material consideration • the council will expect applicants to provide information relating to erosion rates and the threat to sites or buildings. • presumption against developments in Cobbold's Point, Felixstowe until defences are implemented or studies show the threat is not imminent. • development not permitted on grounds of prematurity at Dunwich and Thorpeness until coastal studies are carried out.
Thames Estuary (Suffolk – North Kent)	Variable rates of marine erosion of the 25–45m high London Clay cliffs has resulted in contrasting landslide systems comprising deep–seated rotational slides (erosion exceed weathering), mudslides (erosion in balance with weathering) and shallow part successive rotational part translational slides (free degradation with zero erosion).	Suffolk	Yes	See above. **Suffolk Coastal DC;** see above
		Essex	No	No specific structure plan policies. **Maldon DC;** local plan policy; • applications for development in recognised areas of instability should normally be accompanied by a stability report or statement that the site is stable. Any report should indicate how instability will be overcome and show how development will not endanger people, buildings or adjoining land. **Southend–on–Sea BC;** not considered a significant planning issue. **Castle Point BC;** erosion not considered a planning issue. **Thurrock BC;** no specific local plan policies.
		Kent	None	No specific structure plan policies. **Canterbury City Council;** local plan includes restrictions in two areas; East Cliff and Studd Hill: • the council will safeguard the defined coastal protection zones from development. In unprotected areas the cliff top is in public open space and building would not be permitted, in line with conservation policies. **Swale BC;** Leysdown and Warden Bay local plan includes a development line beyond which no new development will be permitted. The line was established in 1982.

Table 2.2 (cont ...).

AREA OF COASTAL INSTABILITY	GENERAL DESCRIPTION	COUNTIES	STRATEGIC POLICIES	COMMENT, INCLUDING SELECTED DISTRICT LOCAL PLAN POLICIES
Channel Coast	Folkestone Warren; large multiple rotational landslide complex developed in Gault Clay and overlying Chalk. Hythe–Sandgate; large slides developed in Lower Greensand Sandgate Beds. Fairlight Glen; landslide complex developed in Lower Cretaceous Ashdown Sands and Fairlight Clay, comprising mudslides and deep–seated rotational failures. Cliff recession has been around 1m/yr over the last century, with the loss of the famous Lover's Seat. Fairlight Village; rapid cliff retreat through rockfalls affecting cliff top properties. Hastings; rotational slides and rockfalls resulting in damage to local properties. Beachy Head to Brighton; rockfalls off the chalk cliffs.	Kent	None	No specific structure plan policies. **Shepway DC**; District local plan policy: ● planning permission within defined areas of Sandgate will not be granted until a soil survey clearly demonstrates that the site can be safely developed and that the proposed development will not have an adverse effect on the landslide as a whole.
		East Sussex	None	No specific structure plan policies. **Rother DC**; in response to the problems at Fairlight Village the council have formulated a specific policy: ● development not normally allowed in an area south and east of Sea Road, (to be reviewed for incorporation in the 1994 plan). The policy has recently been supported on appeal. **Hastings BC**; no formal policies, although conditions may be attached to planning permissions requiring adequate site investigation and soil report. Work should not proceed until and unless measures deemed necessary by the authority have been incorporated in the development proposals. **Eastbourne BC**; no formal policies.
Isle of Wight Undercliff	Large landslide complex comprising multiple rotational slides with shear surfaces in the Gault Clay, compound slides with basal shear surfaces in the Lower Greensand, large rockfalls off the Upper Greensand rear scarp and mudslides. This landslide has caused considerable disruption to development and infrastructure in the area, particularly in the town of Ventnor, at Luccombe and Blackgang.	Isle of Wight	Yes	Draft structure plan policy: ● development on known unstable land not permitted unless authority can be satisfied that the site can be safely developed and used without adding to instability problems.
Shanklin – Sandown Cliffs	Rockfalls controlled by stress relief, frost action and seepage erosion threatening downslope developments. Currently in process of stabilisation.	Isle of Wight	Yes	Structure plan policies: see above

Table 2.2 (cont ...)

AREA OF COASTAL INSTABILITY	GENERAL DESCRIPTION	COUNTIES	STRATEGIC POLICIES	COMMENT, INCLUDING SELECTED DISTRICT LOCAL PLAN POLICIES
Christchurch Bay	Rapidly retreating Barton Clay Cliffs (1.9m/yr). Lithological variations have been accentuated by landsliding, producing a bench profile. Degradation processes include mudslides, compound failures, rockfalls and debris slides.	Hampshire	Yes	Structure plan policy. ● development will be restricted on those parts of the coast where the forces of coastal erosion exist. **New Forest DC;** no formal policies, although there is a commitment to maintaining and improving existing defences or providing new defences in areas of high risk.
		Dorset	None	No specific structure plan policies. **Christchurch BC;** no local plan policies **Bournemouth BC;** no local plan policies **Poole BC;** no local plan policies; conditions attached to planning permissions require no soakaways within 400m of cliff top.
Weymouth Bay	A wide variety of landslides have been reported along this section of coastline, including landslides in Warbarrow Bay, developed in the Wealden Clay, rockfalls off the Chalk Cliffs, debris slides off the Purbeck and Portland Beds, and compound slides involving the Corallian at Red Cliff.	Dorset	None	No specific structure plan policies. **Purbeck DC;** no local plan policies
Isle of Portland	Large complex landslides, occasionally involving toppling failures as at Great Southwell; ground movement problems along the Undercliff slopes on the north and east of the island.	Dorset	None	No specific structure plan policies. **Weymouth and Portland BC;** instability has been recognised as a constraint, highlighting the need to produce a stability report as part of a planning application
Lyme Bay	A series of massive landslide complexes have developed along this stretch of coast, from Bridport to Sidmouth. These include Fairy Dell, Black Ven and the Landslip Nature Reserve west of Lyme Regis. A wide variety of landslide processes have been recorded, including mudsliding, multiple rotational sliding and block sliding. The characteristic form along this coast is a series of undercliffs which highlight lithological variations. Cliff retreat rates of between 0.4–0.5m/yr were recorded by Brunsden and Jones (1980) at Fairy Dell. In West Bay landsliding has caused considerable problems for coastal development, with rates of cliff retreat of up to 2.8m/yr recorded.	Dorset	None	No specific structure plan policies. **West Dorset DC;** no formal local plan policies.
		Devon	None	No specific structure plan policies **East Devon DC;** no local plan policies, affected areas covered by conservation policies.

Table 2.2 (cont ...)

AREA OF COASTAL INSTABILITY	GENERAL DESCRIPTION	COUNTIES	STRATEGIC POLICIES	COMMENT, INCLUDING SELECTED DISTRICT LOCAL PLAN POLICIES
Budleigh Salterton	A variety of landslide forms occur on the rapidly eroding cliffs west of Budleigh Salterton, including mudslides and rockfalls.	Devon	None	No specific structure plan policies. **East Devon DC**; no local plan policies, affected areas covered by conservation policies.
Torbay	Numerous rockfalls, debris slides and mudslides along the Torbay coastline. There is clear relationship between landslide form and geology (lithology and structure).	Devon	None	No specific structure plan policies. **Torbay BC**; no formal policies, although many areas protected by conservation policies.
Downderry, Cornwall	Rapidly eroding 6–8m high cliffs, affecting cliff top properties. Erosion rate is around 0.1m/yr.	Cornwall	None	No specific structure plan policies. **Caradon BC**; no local plan policies
Bude Bay and Bideford Bay	Rock falls and toppling failures off the hard rock cliffs	Cornwall	None	No specific structure plan policies. **North Cornwall DC**; no local plan policies
		Devon	None	No specific structure plan policies. **Torridge DC**; no local plan policies.
Watchet Bay	Debris slides and rockfalls off the coastal cliffs.	Somerset	None	No specific structure plan policies. **West Somerset DC**; no local plan policies.
Glamorgan Cliffs	Rockfalls, topples and translational slides off the Lias cliffs. Erosion rates of 0.3–0.7m/yr.	South Glamorgan	None	No specific structure plan policies. **Vale of Glamorgan BC**; no specific policies but: • cliff tops designated as a coastal conservation zone; development will not be permitted that does not conform with local plan policies.
		Mid Glamorgan	None	No specific structure plan policies. **Ogwr DC**; no specific policies although cliffs protected by conservation policies.
St Brides Bay	Rockfalls, debris slides and complex landslides developed on the cliffs in St Brides Bay.	Dyfed	None	No specific structure plan policies. **Pembrokeshire Coast National Park Authority**; no formal policies, although conservation policies protect cliffs.
Cardigan Bay	Rockfalls and mudslides developed in low cliffs formed in unconsolidated glacial deposits. Erosion rates up to 0.25m/yr.	Dyfed	None	No specific structure plan policies. **Ceredigion DC**; no local plan policies, although cliffs protected by conservation policies.
The Lleyn Peninsula	Complex landslides and debris slides in superficial deposits.	Gwynedd	None	No specific structure plan policies. **Dwyfor DC**; no specific policies, although: • presumption against planning permission for buildings on cliff tops. This is a conservation policy for the coastal AONB.
Conwy Bay	Complex landslides and debris slides in superficial deposits.	Gwynedd	None	No specific structure plan policies. **Aberconwy DC**; no specific local plan policies

Table 2.2 (cont ...)

AREA OF COASTAL INSTABILITY	GENERAL DESCRIPTION	COUNTIES	STRATEGIC POLICIES	COMMENT, INCLUDING SELECTED DISTRICT LOCAL PLAN POLICIES
Inner Hebrides	Skye; large multiple rotational slides developed in Tertiary basalts overlying Jurassic sedimentary rocks. Examples such as the Storr and Quirang at Trotternish are amongst the largest landslides in Great Britain.	Strathclyde	None	No specific structure plan policies.
	Mull; rockfalls and complex slides of the basalt cliffs of south and west Mull, as in the Wilderness area.	Highland	None	No specific structure plan policies.
	Arran; rockfalls off the New Red (Permian) sandstone of south Arran.			
Fife Coast Cornwall	Rockfalls and rockslides on the Carboniferous rock cliffs. Footpaths have been truncated and historical monuments affected.	Fife	None	No specific structure plan policies.
Tyne to the Tees	Rock falls and slides off the coastal cliffs	Durham	None	No specific structure plan policies.
		Cleveland	None	No specific structure plan policies.
North Yorkshire; Filey to Runswick Bay	This stretch of coastline has a long history of landslide problems, especially in Runswick Bay where the old village slid into the sea in 1689. Nearby Kettleness was lost in 1829. Landsliding problems are also significant at Whitby, Scarborough, Cayton Bay and Filey. The Holbeck Hall landslide of June 1993 was the most recent large event. The main type of reported failure include complex slides, mudslides and debris slides.	North Yorkshire	None	Instability viewed as a local rather than strategic planning matter. **Scarborough BC;** no specific policies, although instability is taken into account when considering individual proposals.
Holderness	Rapidly retreating cliffline with rates of retreat up to 12m/yr, averaging 1.8m/yr. Over the last 1000 years at least 30 villages have been lost to the North Sea. Retreat occurs through a sequence of deep-seated rotational slides, with occasional debris slides and mudslides.	Humberside	None	Policies are likely to be included in the Structure Plan review. **Holderness BC;** no formal local plan policies, although council holds view that planning permission ought not to be granted in erosion risk areas, where rate of erosion was likely to remove development in it's normal life span (ie set-back).

Figure 4.5 The pattern of recorded landslide activity on the south coast of England (above) compared with annual rainfall patterns (below; from the Isle of Wight).

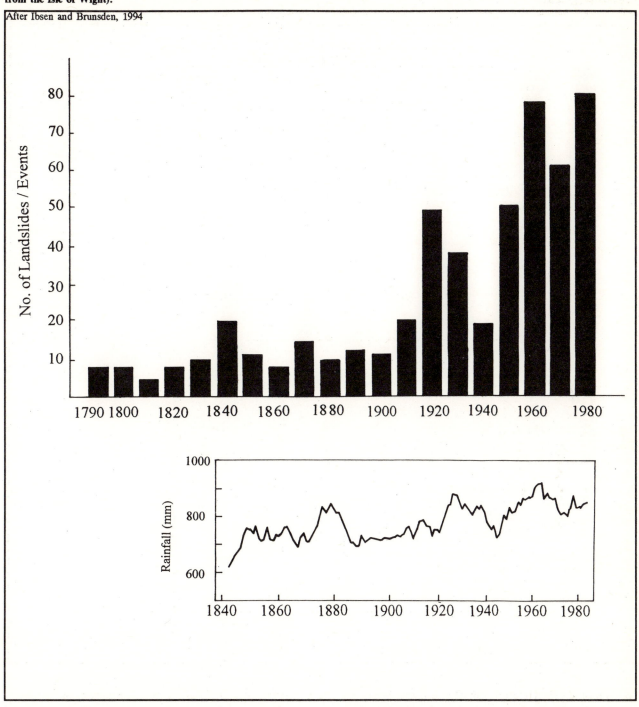

After Ibsen and Brunsden, 1994

shingle ridges to maintain a balance between erosion and accretion, a fully quantified **sediment budget** is often not a practical proposition. Under such circumstances a general indication of the relative significance of sediment movement and those landforms that may be vulnerable to reduced sediment supply may be appropriate; this follows the **precautionary principle** by which a cautious approach is needed where it is suspected that there is potential for significant sediment disruption. In this context, sediment budget considerations should be viewed as necessary to establish the

sustainability of natural coastal defences, conservation sites and recreation areas, and operations such as aggregate extraction and dredging.

Sensitivity

The response of coastal landforms to the effects of natural processes or human activity can vary considerably. For example, a storm passing across

47

Figure 4.6 The relative sensitivity of different landslide systems within the Isle of Wight Undercliff to winter rainfall totals (after Rendel Geotechnics, 1995c).

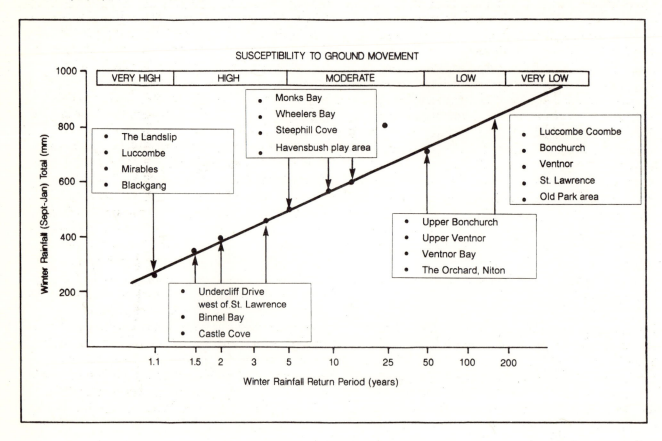

two coastal systems may result in very different patterns of erosion, deposition and flooding. In some areas major storms can have a profound effect on the coastline; on the Grampian coast, for example, storms have let to the formation of bars across harbour mouths and the shifting of sand dunes. Elsewhere, there may be no obvious effects. This complexity of response is a measure of the **sensitivity** of a coastline which can range from:

(i)   fast responding systems which are very sensitive to disturbing events. This type of system can be morphologically complex because the landforms are subject to rapid change, such as sand dunes;

(ii)   slowly responding insensitive systems, such as hard rock cliffs.

The main controls of sensitivity include material strength, morphological resistance (e.g. relative relief, altitude, slope angle) and the importance of linkages between landforms within a system (i.e. sediment budget). A stable coast, for example, is one in which the controlling resistances are sufficient to prevent a storm event from having any effect. Thus, although wind and wave energy is

greatest on the north and west coasts this does not correspond with rapid erosion and sediment transport.

The sensitivity of a coast to change can be considered at a range of scales. At a general level it is possible to recognise coastal systems where the large amounts of sediment transport and the strong long-shore links between the different landforms indicate **high sensitivity**. Other coasts may be characterised by little sediment transport and can be considered to be of **low sensitivity**, (Figure 4.3). In general, the more sensitive coastlines appear to be associated with soft sedimentary rocks and glacial materials, in areas where relatively strong longshore currents occur i.e. much of east and south England from Flamborough Head to Dawlish Warren, parts of west and north Wales and the Solway Firth.

Within coastal systems there may be notable variations in sensitivity, reflecting the nature and composition of individual landforms or changes in the degree of exposure to waves and tidal currents. Amongst the most sensitive elements of a coast are often sand dunes, beaches, shingle ridges, mudflats and saltmarshes. Many storms are associated with widespread foreshore erosion and offshore accretion as beach or mudflat sediment are redistributed in the face of very high wave energies. Although permanent changes can result, much of the "damage" to these achieved by these events can be restored over the forthcoming months and years, as sediment is transported back onshore by wave and tidal energy. The 1953 storm surge, for example, caused extensive erosion: sand and shingle were stripped from the beaches, dunes were undercut and marshes flooded. After the storm, however, there was considerable accretion on offshore bars which developed as a result of the storm moved inshore. Some areas quickly gained more material than they had lost (Barnes and King, 1953). The recovery can be hindered, however, if the supply of sediment to these features is disrupted by, for example, groynes, coast protection works or harbour walls. This may lead to a trend of net erosion.

Estuaries, for example, can be very sensitive to the effects of changing patterns of erosion and accretion, with mudflats and saltmarshes effective in dissipating tidal and wave energy and, hence, reducing the risk of flood. Theoretically, the predicted increases in sea level could be balanced by continued deposition throughout the estuary, so that the growth of mudflats and saltmarshes keeps pace with average water levels. However, in order to keep pace with the predicated increase in mean sea levels many estuaries would need to accrete sediment at a rate in excess of predicated rates of sea level rise of between 6mm – 15mm a year by 2030 (IPCC, 1990). Any future shortfalls in the rate of sediment supply to some estuaries could affect the level and extent of mudflats and saltmarshes and lead to an increase in flood risk.

Estuaries can also be sensitive to the effects of the progressive control of the channel through the construction of embankments and walls. These defences can lead to a reduction in the floodwater storage capacity of estuarine lowlands and marshes. Horner (1978) has demonstrated that high tides on the Thames at London Bridge have risen over the last 150 years (Figure 4.4a); a similar trend has been identified in the Humber (Humberside County Council, 1994; Figure 4.4b). These rises can only be partly explained by rising sea levels and it is now believed that confining the tide to a narrow channel between high flood embankments may lead to a rise in the tidal range and increase the risk of flooding.

Although change on coastal clifflines is **progressive**, involving retreat rather than recovery,

many cliffs undergo cycles of instability with erosion events (e.g. rockfalls, landslides, etc.) followed by periods of relative stability during which the cliff is "prepared" for further failure. In this context many cliffs are sensitive to changes in the intensity of wave attack at the cliff foot or variations in groundwater levels within the slope; both factors can be influenced by human activity. Figure 4.5, for example, compares the frequency of recorded landslide activity on the south coast of England from 1840 with variations in the annual rainfall totals over the same period, and demonstrates a broad association of landsliding with wet years. Indeed, this pattern serves to demonstrate that climate is not constant, leading to **uncertainty** with regard the occurrence of coastal landslides and cliff recession. The relationship between landslide activity and rainfall is not a simple one. Some coastal cliffs and landslide systems are particularly sensitive to rainfall events whilst others appear only to show signs of movement during extremely rare conditions. An example of the differences in sensitivity to rainfall can be seen in Figure 4.6 which compares the return period of different winter rainfall totals with the known history of events in different landslide systems within the Isle of Wight Undercliff, since 1839. This suggests that:

- periods of accelerated ground movement have not occurred in Ventnor and St. Lawrence during winter rainfall conditions that can be expected one year in 100 or more;

- major movements at Luccombe Village have occurred during conditions that can occur one year in four;

- major movements in The Landslip and near Mirables have occurred in response to winter rainfall totals that can be expected around one year in 1.5 or less;

Sensitivity is an important consideration in coastal planning and management as it can provide an indication of the susceptibility to changes associated with sea level rise and the effects of development. These changes can be manifest through modifications to the degree of risk and the environmental quality of the coast.

5 Methods and Techniques for Data Acquisition

Introduction

The three key considerations – risks, sediment budget and sensitivity – are relevant on all coasts and at all stages of the decision making process, from regional or strategic level to the determination of site specific proposals. In Chapter 3 an hierarchical model was set out which indicated the investigation approach suited to different stages of the decision–making process; from general assessment to site reconnaissance and detailed assessment. As the information required for these investigations changes from a general awareness of the coastal environment to the need for site specific information, the questions that should be considered remain broadly the same (Figure 5.1). The significance of some questions and the extent to which they will need to be addressed will, however, vary according to the nature of the coastal environment and the extent to which it is subject to pressure for development. These variations are examined in Chapters 6–9 with specific reference to the information needs in coastal lowlands, estuaries, sand dunes and cliffs.

Before examining specific information requirements in different environments, it is necessary to consider what types of earth science information are needed to address the key considerations and the ways in which such information can be integrated, analysed and presented to answer the specific questions posed by planners and managers. Given the wide variety of coastal environments and associated issues it is inappropriate to be dogmatic about the precise methods that can be employed in data collection exercises. However, in most cases an understanding of **landforms** is a key component of most investigations. This is because landforms are capable of study at a wide range of scales from individual sites to major coastal systems and can often be a sensitive indicator of **materials** and **process.**

The identification and mapping of landforms can provide a spatial framework for understanding site specific information on materials or processes and for extrapolating this information into less well investigated areas. Recognition of particular landforms can also be fundamental to the assessment and prediction of potential environmental change. Indeed, as decision–makers often require immediate knowledge of potential changes without giving opportunity for long–term monitoring, it is frequently necessary to estimate change and processes through the interpretation or modelling of landforms and materials.

However without exception, an **inventory** of landforms, alone, will not provide the right documents to pass onto planners and managers. This information will need to be **interpreted** in terms of risks, sediment budget or sensitivity, to show features specifically relevant to the management concerns in that area. Thus, for example, management of urban landslide problems may require an assessment of the landslide hazard and the potential consequences of ground movement in map form. Other data, if shown on the maps, may only serve to confuse or divert attention away from the main issues facing planners and managers.

In many instances, it may be difficult for non-technical users to make judgments about the relative significance of information shown on different interpretive maps (ie. risks, sediment budget, sensitivity). Consequently, a series of these derived maps will often need to be integrated and **summarised** to indicate, in general terms, the principal constraints or resources in an area. The end product, essentially a form of **sieve map**, should be familiar to and accessible by planners and managers, ensuring that maximum value is gained from the earth science information. Such maps also provide a mechanism for ensuring that the key environmental parameters are expressed in terms of their significance for planning and

Figure 5.1 Elements of earth science investigations.

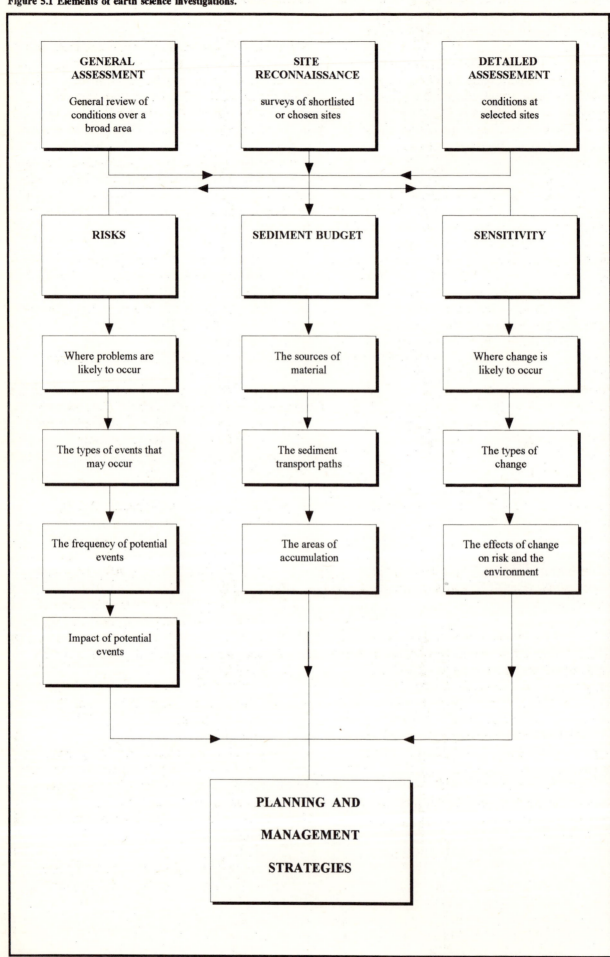

development. For example, **planning guidance** can be prepared to highlight areas that are suitable for development, have significant constraints or are largely unsuitable for specific types of development on the basis of the risks or the sensitive nature of the environment. Where there are concerns about the possible environmental effects of a proposed development, the collected and interpreted information may need to be summarised by the developer in terms of an **environmental assessment**.

The following sections are intended to outline the general principles of the investigation procedure from **inventory** to **interpretation** and **summary** (Figure 5.2). The 5 trial study investigations presented in Annex A are intended to support this section by providing examples of the way specific issues in different coastal environments can be addressed through these procedures (Figure 5.3). Owing to resource constraints these examples address the information needs for planners and developers at a broad, strategic level, suitable for providing an input into the development plan process. The procedures are, however, relevant at each of the 3 levels of investigation identified in Chapter 3 (Figure 5.1). The approach will also have wider applicability to other bodies with an interest in aspects of coastal management, including coastal defence operating authorities and groups, and conservation agencies (Figure 5.4). Some elements of the approach will, however, become more important for particular end-users. For example, planners are generally best served by summary maps, whereas the technical officers within, for example, a coast protection authority would gain greater benefit from the interpretative or inventory maps.

Before proceeding further it is important to stress that lack of awareness of the key issues in an area is often a significant obstacle to be overcome, especially where risks are associated with extremely rare events of the failure of man-made structures. It is important, therefore, that the first step in any investigation should be to establish what type of issues could be expected in particular coastal environments (Table 5.1; Figure 5.5). Here, advice should be sought from local authority engineers or the NRA about whether these issues are likely to give rise to significant planning considerations.

Inventories

A first step in most data collection exercises will be to produce an inventory of coastal features and to characterise these features in terms of materials, form and processes. The best systematic and nation-wide survey of coastal landforms was produced by the then Nature Conservancy Council as part of its Atlas of Coastal Sites Sensitive to Oil Pollution (see Appendix A.2.1). However the maps, at 1:100,000 scale, provide only a broad indication of the coastal features and for many studies it will be necessary to seek more detailed information from:

- Ordnance Survey maps (Appendix A.5.1);
- Hydrographic charts (Appendix A.5.3);
- Aerial photographs (Appendix A.5.4);
- The AA Illustrated Guide to Britain's Coast (Appendix A.2.1);
- Field inspection;
- Consultation with key interest groups.

Individual landforms or groups of landforms should then be **characterised** in terms of their materials and processes, with reference to available sources of information or, if necessary, specially commissioned studies. Amongst the most important data sources will generally be:

- Geological maps (Appendix A.2.2);
- Soil maps (Appendix A.2.2);
- National Reviews of Ground-Related Problems (Appendix A.2.2);
- The UK Digital Marine Atlas (Appendix A.2.3);
- The Macro-Review of the Coastline of England and Wales (Appendix A.2.3);
- The Beaches of Scotland (Appendix A.2.3);
- Shoreline management plans (Appendix A.2.3);
- Historical records (Appendix A.5.5);
- Sources identified by the Coastal Zone Database (Appendix A.1)

The identification and characterisation of landforms provides an essential basis for understanding the nature of a particular coastline. The information collected is often best presented as a **Geomorphological Map** portraying some or all of the following;

(i)   **Surface Form;** morphology and slope angle (Figure 5.6), offshore bathymetry;

Figure 5.2 The investigation procedure.

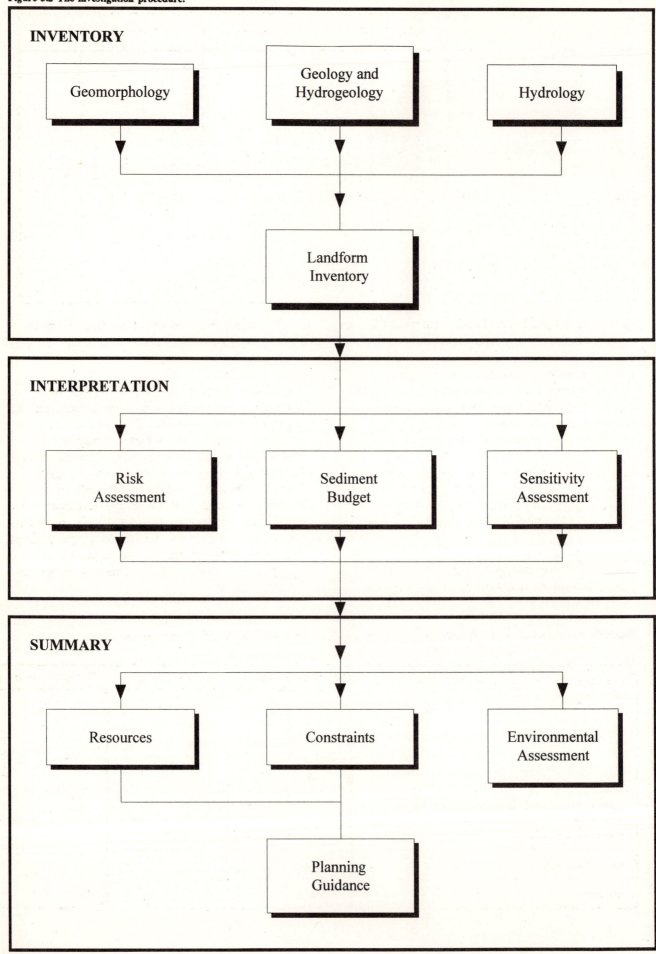

Figure 5.3 The 5 trial study areas: Summary of characteristics.

TRIAL STUDY AREA	Coastal Types				Environments				Maps Produced					
	Developed	Despoiled	Undeveloped - Conserved	Undeveloped	Estuaries	Coastal Lowlands	Coastal Cliffs	Sand Dunes	Geology and Geomorphology	Risks	Sediment Budget	Sensitivity	Resources	Constraints
Fraserburgh - Scotstown Head	•		•	•	•	•		•	•		•		•	•
Seaham - Teesmouth	•	•	•	•	•	•	•	•	•			•	•	•
Filey - Scarborough	•		•	•			•		•	•			•	•
Flamborough Head - Humber Estuary	•		•	•	•		•	•	•		•		•	•
Cromer - Sea Palling	•		•			•	•	•	•			•		•

(ii) **Geological Conditions;** bedrock lithology, superficial deposits, structural form;

(iii) **Processes;** generally surface form or materials can be mapped to indicate the dominant process. For example:

- **erosion** can be portrayed by the presence of landslides or eroding cliffs;
- **flooding** can be mapped by identifying the extent of land affected by past floods.

The map can also indicate the nature of **sediment transport processes** by highlighting sediment sources and sinks, together with dominant directions of littoral drift.

A geomorphological map will seldom be an appropriate end product to pass on to planners and managers. Such maps will often appear too complex and many of its features may be meaningless to the non-specialist. The map may also carry data which are not relevant to his immediate concerns and the significance of the various map units to planning and development may be unclear. For many coasts, decision-makers will be better served by specifically prepared maps, based on an interpretation of the inventory information, to address **risk, sediment budget** or **sensitivity** issues.

Figure 5.4 Bodies with an interest in earth science information collection at different levels of the investigation procedure.

* Operating Authorities include : the NRA, IDBs and Local Authorities	General Assessment			Site Reconnaissance			Detailed Assessment		
	Risks	Sediment Budget	Sensitivity	Risks	Sediment Budget	Sensitivity	Risks	Sediment Budget	Sensitivity
Coastal defence Operating Authority* and Groups	•	•	•	•	•	•	•	•	•
Local Planning Authority	•	•	•	•	•	•			
Conservation Agencies			•			•			•
Developers				•	•	•	•	•	•

Table 5.1 A selection of key issues associated with earth sciences in different coastal environments.

Coastal Environment	Risk Issues	Sediment Budget Issues	Sensitivity Issues
Estuaries	• Flooding • Sedimentation • Channel Erosion • Problem Foundation Conditions	• Sedimentation in watercourses, increased flood risk and navigation problems • Disposal of dredgings on land or at sea • Maintenance of natural coastal defences e.g. mudflats and saltmarshes • Effects of mineral extraction from the foreshore on erosion and flood risk • Effects of mineral stockpiles on floodplain storage	• Creation and maintenance of conservation sites • Effect of channel maintenance, flood defence and mineral extraction on conservation sites • Effect of sea level rise on flood risk
Coastal Lowlands	• Flooding	• Maintenance of natural coastal defences, e.g. mudflats, saltmarshes, beaches, sand dunes • Maintenance of amenity beaches and sand dunes • Effects of mineral extraction from the foreshore on erosion and flood risk	• Creation and maintenance of conservation sites • Effects of flood defence on conservation sites • Effect of sea level rise on flood risk
Coastal Cliffs	• Landsliding • Cliff Recession	• Supply of sediment to natural coastal defences e.g. mudflats, saltmarshes, beaches, sand dunes • Supply of sediment to amenity beaches and sand dunes • Effects of mineral extraction on erosion and flood risk	• Creation and maintenance of conservation sites • Effects of coast protection on conservation sites • Effect of sea level rise on erosion rates
Sand Dunes	• Wind Blown Sand • Flooding	• Supply of sediment and maintenance of natural coastal defences, e.g. beaches • Supply of sediment and maintenance of amenity beaches and sand dunes • Effects of mineral extraction on erosion and flood risk	• Creation and maintenance of conservation sites • Effects of coast protection and mineral extraction on conservation sites • Effects of sea level rise on erosion rates

Risk Assessment

The terminology used when discussing natural events and their impact can be misleading. It is not uncommon for terms like hazard, risk and vulnerability to be used interchangeably by different people. Here, the following definitions are used (after Varnes, 1984):

• **hazard** describes the chance of a potentially damaging event occurring within an area;

• **risk** describes the possible losses arising as a result of a damaging event;

• **vulnerability** describes the degree of loss or damage to particular sectors of a community (e.g. buildings, infrastructure, services etc.) at risk from an event i.e. different elements will face different levels of risk depending on their vulnerability.

In most instances risk assessment will involve some or all of the following steps:

Figure 5.5 Coastal environments.

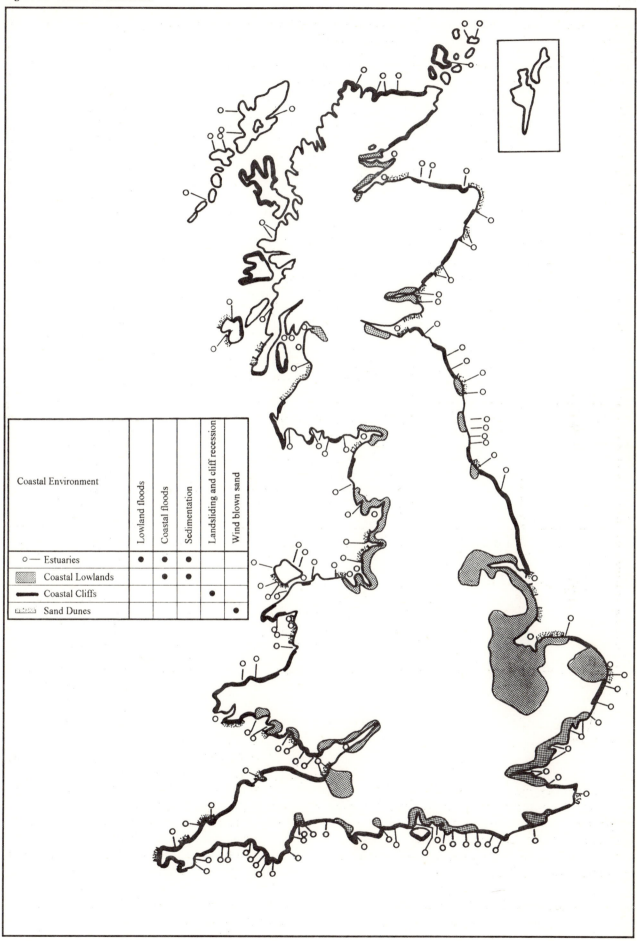

Coastal Environment	Lowland floods	Coastal floods	Sedimentation	Landsliding and cliff recession	Wind blown sand
○— Estuaries	●	●	●		
Coastal Lowlands		●	●		
Coastal Cliffs				●	
Sand Dunes					●

Figure 5.6 Morphological and geomorphological mapping (above; after Cooke and Doornkamp, 1990, below; after Lee and Moore, 1989).

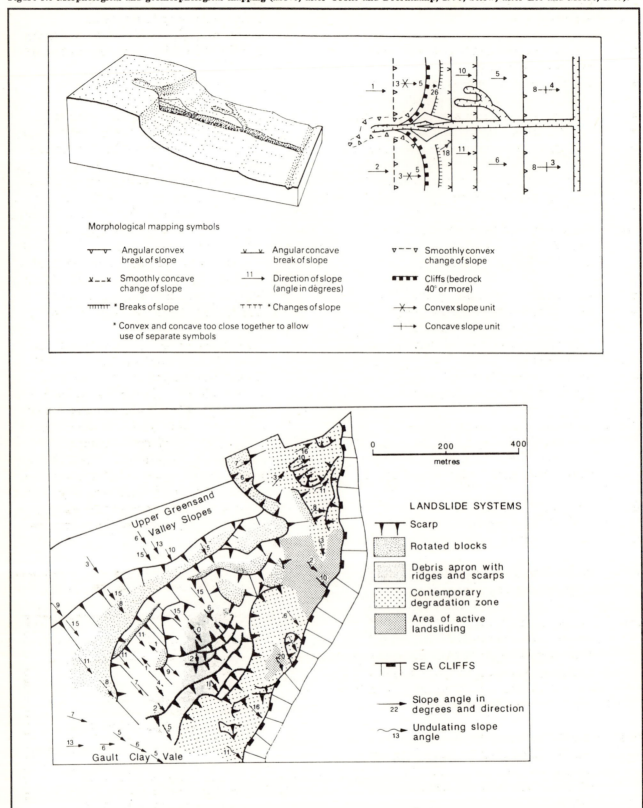

(i) **defining hazard zones** which could be affected by erosion, landsliding, flooding etc. This can be achieved by mapping surface form or topography to delineate the extent of vulnerable land;

(ii) **defining hazard potential** in terms of the nature and size of events that could be expected in an area. This will require an appreciation of:

- the nature and magnitude of historical events (Appendix A.5.5);
- the theoretical occurrence of events from rainfall or tide and wave climate records;
- causes and mechanisms of possible events;
- factors influencing the pattern of events;
- the effects of development on the incidence of events;
- the standard of protection provided by existing coastal defences

(iii) **identifying land use elements at risk;** land use within hazard zones can be classified according to:

- the **relative importance** (Table 5.2)
- the **vulnerability of the buildings** expressed in terms of structural vulnerability and material vulnerability (eg. Table 5.3)
- the **long term effects** to agricultural land from flooding

(iv) **assessing the potential consequences;** the severity of the potential consequences of each type of event can be considered in terms of a simple 3-fold classification:

- **total loss;** ie. deaths, severe casualties or destruction of property
- **partial loss;** ie. injury, minor casualties or serious and severe property damage;
- **minor loss;** ie. inconvenience or slight and moderate property damage

(v) **assessing the likelihood of damaging events;** an appreciation of the size and frequency of potentially damaging events is a central component of risk assessment, and will generally be expressed as a **return period** or **recurrence interval.**

(vi) **calculating annual risk and reliability;** standard expression for calculating the specific risk associated with an event of particular magnitude is calculated as follows:

Likelihood of an Event
Importance of Land Use x Vulnerability of Structures

The measure of chance that an engineering system or structure will not fail is known as the **reliability**. From a coast erosion perspective, it can be viewed as an indication of the ability of a slope to support particular land uses or development. Reliability is related to the probability of failure, i.e. risk, as follows:

Reliability = 1 − Risk

In many instances it is inappropriate to evaluate risk or reliability in absolute terms because of the uncertainties in assigning values for the hazard and the assets at risk. In many cases it is more useful to assess the **relative risk** to each of the shortlisted sites from particular hazards, based on both factual data and subjective appraisal. The value of relative risk assessment is that it can quickly enable sites to be compared or allow decisions to be made about where limited resources and finances should be directed (Rendel Geotechnics, 1995b).

An example of the use of a relative risk approach is provided by Clark et al (1993) for the management of weak sandstone cliffs at Shanklin, Isle of Wight. Here potential cliff face and talus slope failures were evaluated by ground inspection of a series of cliff sections, enabling a systematic assessment to be made of the hazard along the entire cliff sections. The risk posed by the failures was established by assigning estimated hazard values to each of the cliff sections. These sections

Table 5.2 Typical values of importance of land use and importance of structures and services (after Cole et al, 1993).

	Land Use Category	Value
1.	Public open space, farmland, tidal land.	0.3
2.	Domestic houses (single family occupancy), secondary communications network/roads and railways, small factories and small places of assembly.	1
		3
3.	Domestic multiple occupancy, places of assembly, medium to large factories and offices, main roads and railways.	10
4.	Essential services, valuable and/or costly property.	30
5.	Structures or services giving great danger if damaged.	

Table 7.2 Estuaries: Key sources of information and advice for general assessments.

Key considerations	England and Wales	Scotland	Sources of Specialist Advice
Risks: Tidal Flooding	• NRA maps of areas prone to flooding (S.105 Maps) • NRA Sea Defence Survey	Various possible sources, including: • Island and Regional Council Water Services Depts. or equivalent • River Purification Authority • Ordnance Survey maps can be used to identify the extent of land below 5m	NRA; Scottish Regional Councils; Coastal Defence Groups; Proudman Oceanographic Laboratory; Storm Tide Warning Service of the Meteorological Office.
Risks: Problem Ground Conditions	• British Geological Survey maps and memoirs • British Geological Survey borehole records • Soil survey maps • historical maps and records • Review of Foundation Conditions (Wimpey, 1994)	• British Geological Survey maps and memoirs • British Geological Survey borehole records • Soil survey maps • historical maps and records • Review of Foundation Conditions (Wimpey, 1994)	British Geological Surveys; Geotechnical consultants.
Sediment budget	• Estuary Management Plans • Shoreline Management Plans • British Geological Survey maps and memoirs • Macro–Review of the Coastline (HR Wallingford)	• Estuary Management Plans • Shoreline Management Plans • British Geological Survey maps and memoirs • The Beaches of Scotland	NRA; Scottish Regional Councils; Coastal Defence Groups; Coast protection authorities; Conservation agencies.
Sensitivity	• Estuary Management Plans • Shoreline Management Plans • Water Level Management Plans • Coastal Management Plans • Conservation agency maps and records • NCC Atlas of Coastal Sites Sensitive to Oil Pollution	• Estuary Management Plans • Shoreline Management Plans • Conservation agency maps and records • NCC Atlas of Coastal Sites Sensitive to Oil Pollution	NRA; Scottish Regional Councils; Coastal Defence Groups; Conservation agencies; Coast protection authorities.

various groups with an interest in the use of the estuary. In this respect, local planning authorities should make full use of the information which these groups are collecting as part of the preparation and review of **estuary management plans** and **shoreline management plans.**

Information Needs for Developers

Site reconnaissance surveys (ie. preliminary assessment), involving a compilation of available geological information (Appendix A.2.2) and ground inspection, should be undertaken to establish whether shortlisted sites are likely to be affected by problem ground conditions. The results of **general assessments**, if publicly available, may provide access to relevant information and alert

developers to the presence to potential problems.

Where there are reasons for suspecting problems ground conditions, the developer should undertake appropriate site investigations and geotechnical appraisal to establish whether:

- the land is capable of supporting the loads to be imposed;
- the development will be affected by contaminated land problems;
- the development will initiate land instability problems on adjacent land.

This report is not the place for a detailed description of site investigation procedures; attention is drawn, however, to the following important publications:

- BS5930 Code of Practice for Site Investigations (BSI, 1981);
- CIRIA Site Investigation Manual (Weltman and Head, 1983).
- Site Investigation in Construction: 1 – without site investigation ground is a hazard (Site Investigation Steering Group, 1993).

In many estuaries it may be appropriate for a developer to identify, at an early stage of preparing a development proposal, the key environmental concerns that could arise. A **scoping study** (see Chapter 5) could provide a general indication of the possible impacts and mitigation measures, through consultation with key interest groups and authorities. Formal environmental assessment may be required for specific types of project (see Table 2.1).

8 Information Needs: Sand Dunes

Introduction

Coastal dunes cover an estimated 56,000ha around Great Britain (Figure 8.1). They are accumulations of wind blown sand, carried inland from a beach plain exposed at low water, mainly during the period of sea level rise that followed the end of the last glaciation. The Sefton dunes, for example, are believed to have developed in a main phase of dune building between 4600–4000 years ago (Plater et al, 1991; Innes and Tooley, 1991), derived from foreshore and nearshore sand deposits. Specialised plants (eg. maram grass and sea lyme–grass) help trap the sand. The extent to which mobile dunes develop into stable features depends on a variety of factors including: sand supply, beach profile and the prevailing wind and wave conditions.

Five main types of coastal dunes can be recognised around the British Coast (Ranwell and Boar, 1986; Figure 8.2):

(i) **Offshore island dunes**; developed on barrier islands as linear features reflecting the direction of longshore drift. Examples include Blakeney, Norfolk and Morrich more (Ross and Cromarty);

(ii) **Prograding dunes**; formed on an open coast where there is an abundant supply of sand either from longshore drift in two directions or from a very shallow sandy shore. Examples include Winterton Ness, Norfolk and Barry Links (Angus);

(iii) **Spit dunes**; formed on sandy promontories at the mouths of estuaries. Examples include Studland dunes, Dorset and the Sands of Forvie, Grampian.

(iv) **Bay dunes**; developed in bays within the shelter of rock headlands, forming a half–moon shape in the beach and outer dune zone. They are characteristic of the rocky indented coasts of south west England, Pembrokeshire and Scotland;

(v) **Hindshore dunes**; formed on extensive sandy coasts where the prevailing wind is onshore, driving sand inland for considerable distances as a series of dune ridges or mobile parabolic dunes. Examples include Braunton Burrows in Devon, Newborough Warren in Anglesey and Culbin Sands in Grampian. This type of dune is also known as the **machair**, characteristic of the western and northern coasts of Scotland.

Dry, loose sand is considerably more vulnerable to wind erosion than damp sand. Problems can occur, therefore, when the seaward faces of dunes are eroded by wave action, exposing the core of the dune. Extremely low tides or periods of relatively low sea level may expose large areas of loose sand which, if dried by strong winds, can be a source of blown sand. The wind speed and direction are critical for determining the erosivity of a particular storm. Whilst sand can be moved by relatively light winds, significant events are probably associated with severe storms with wind speeds of up to 100 knots or more (over 12 on Beaufort scale).

Migration of sand dunes is not a significant problem in Great Britain, but it once was. There is

Figure 8.1 Coastal dune systems (after Doody, 1989).

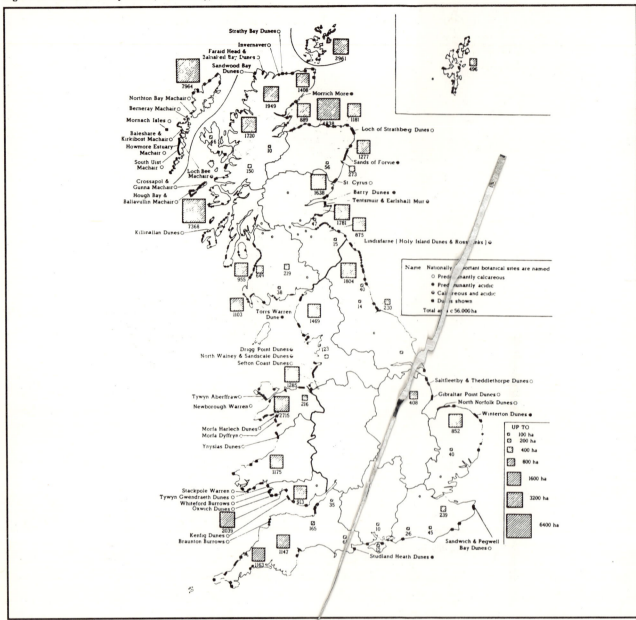

considerable historical evidence to suggest that wind blown sand was a major hazard, especially during the 15th to 17th centuries. Perhaps the most memorable series of events was the Culbin Sands disaster of 1694 and following years. At that time, 16 fertile farms covering some 20–30km² near Findhorn and Forres on the Moray Firth were overwhelmed in a single violent storm. The whole area including the mansion house was buried by up to 30m of loose sand. From 1694 to 1704 there were frequent periods of blowing sand and the area remained a desert of shifting sand for 230 years until it was successfully afforested by the Forestry Commission during the 1920s.

Blown sand problems do still occur, but tend to be localised and considerably less dramatic. The best

known sites include Braunton Burrows and Sefton in England, and the mobile sand sheets at Balmedie, Foneran and Forvie in Scotland. At Braunton, for example, the central dunes migrated 122m inland between 1885 and 1957, although they have now been largely stabilised by vegetation. In the early 1900's mobile dunes threatened the Liverpool–Southport railway at Sefton (Jones et al, 1991).

Land management is a major factor in both preventing and initiating wind–blown sand problems. Indeed, afforestation of the Culbin Sands, Morrich More and other mobile sand sheets has proved successful in stabilising the dune systems. Elsewhere the planting of marram grass has reduced the local problems associated with

Figure 8.2 Major types of dune systems (after Ranwell and Boar, 1986).

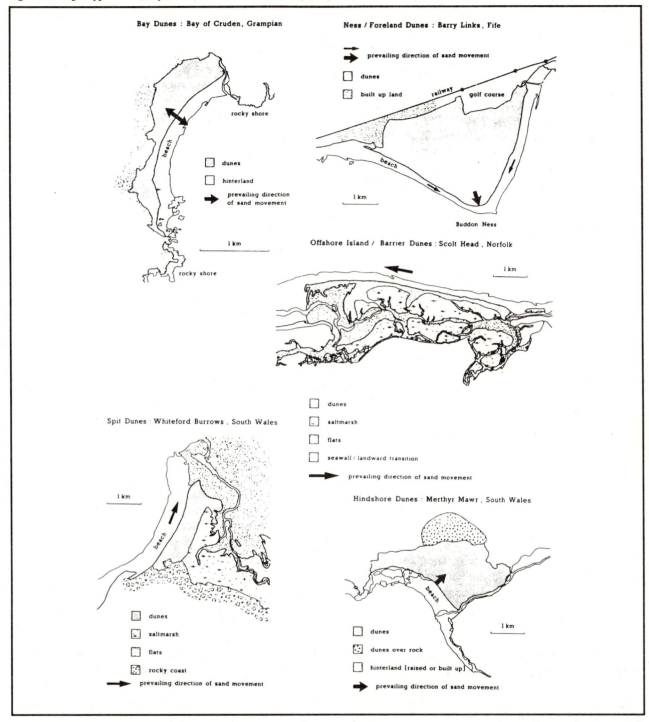

wind blow that occur in many dune systems. The importance of these stabilisation measures is reflected by the widely recognised dangers of uncontrolled marram-cutting; legislation was passed in Scotland after the Culbin Sands disaster and in 1742 an Act of Parliament was laid down "for the more effectual preventing of the cutting of Star or Bent".

Recreational use can also lead to destabilisation of dunes. Indeed, concern about the impact of recreation led the Countryside Commission for Scotland to commission a survey of all the beach and dune systems in Scotland (Ritchie and Mather, 1984). Particular problems can be caused by:

- access to beaches from car parks
- holiday accommodation
- trampling of dunes
- off road vehicles

On exposed west coast sites the effects can be severe with major hollows or "blow-outs" and loss of surface vegetation and sand during storms.

Rabbit burrowing, associated with heavy grazing by cattle and recreational use, can destabilise dunes. In some machair areas of west coast Scotland this has lead to excessive sand movement (Bond, 1978). Golf courses represent one of the major uses of sand dunes in Britain. Over a third of all dunes have been partly modified for this activity (Ranwell, 1975). The positive management of dune areas, with the encouragement of fairway and "rough" grasses has tended to reduce wind blow problems and provide suitable conditions for the survival of dune vegetation.

The consequences of erosion can be significant as many dunes have an important role in providing "natural" defence against tidal flooding, by forming a barrier of high ground in front of a low flying coastal plain, as at Harlech in west Wales. In addition, during severe storms marine erosion of the seaward dunes can mobilise significant quantities of sediment, much of which may accumulate on the foreshore where it helps dissipate wave energy and, hence, reduce the potential for further erosion. Following the storm, much of the "damage" to the dunes will be repaired naturally, as sand is blown back off the foreshore.

Sand dunes can be highly sensitive to changes in the supply and transport of sediment and, hence, particularly vulnerable to the effects of coastal development. Of particular significance, disruption of longshore sand transport may lead to the eventual deterioration of the dunes and increase the likelihood of breaching and, hence, the risk of flooding. Dunes and fronting beaches may contain valuable sand and gravel resources. However, their removal for the construction industry can adversely affect the rate of coastal erosion, and the vulnerability to flooding of adjacent land (see PPG 20, DoE, 1992a). In Scotland, large scale sand extraction is common on the dune and beach complexes of the Orkney and Shetland Islands (eg., the Bay of Quendale) and, to a lesser extent, around Brodick and Girven on the Clyde coast. It has been estimated that sand extraction has had an adverse effect on 16% of all beach complexes in Scotland (Ritchie and Mather, 1984).

Sand dunes are of major conservation value, for both the natural habitats and important geomorphological features. Sand dunes managers, however, recognise the importance of a degree of wind erosion and instability, together with marine erosion of the seaward dunes, to maintain a full sequence of successful stages of habitats. In this context, change is a natural process with the stability imposed by some dune management techniques acting to preserve rare species rather than encouraging successional change.

It should be noted that sand dune migration is principally a land management issue, rather than a planning and development issue. However, planners should be aware of the important role that sand dunes may have in providing natural coastal defences to low lying areas inland. Maintenance of dune systems can, therefore, be an essential component of a coastal defence strategy. Planners should, therefore, ensure that development within the dunes or on neighbouring stretches of coastline does not reduce their effectiveness through, for example, inappropriate mineral extraction or the disruption of coastal sediment transport. The key planning considerations for sand dune areas will, therefore, include:

(i)   **risks**; the possible breaching of sand dune areas and flooding of lower lying land behind (see Chapter 6); localised wind blown sand problems;

(ii)   **sediment budget**; the effects of mineral extraction on the sustainability of natural coastal defences and recreational features;

(iii) **sensitivity**; the permanent or temporary effects of storm conditions on the standard of protection provided by natural coastal defences; the potential effects of the sea level rise on the level of risk; the sensitivity of conservation features to the effects of development.

These considerations have been examined in the following trial study areas (see Annex A):

- Fraserburgh – Scotstown Head, Grampian.
- Seaham – Teesmouth, Cleveland and Durham;
- Cromer – Sea Palling, Norfolk;

The information needs for assessing flood risk considerations have been outlined in Chapter 6, to which reference should be made. As wind blown sand does not constitute a significant hazard around the British coast, this Chapter concentrates, therefore, on examining the information required to address the minerals planning and conservation issues that are frequently associated with sand dune environments.

Background Information

Basic information sources relevant to coastal sand dunes have been identified and compiled in the Coastal Zone Database (Appendix A.1). The most important will generally include:

(i)   **landforms** (Appendix A.2.1):

- the NCC Atlas of Coastal Sites Sensitive to Oil Pollution;
- Aerial photographs.

(ii)   **materials**; (Appendix A.2.2):

- British Geological Survey maps and memoirs;
- Soil maps and memoirs;

(iii)   **processes**; (Appendix A.2.3):

- Shoreline management plans
- Macro-Review of the Coastline of England and Wales
- The Beaches of Scotland

(iv)   **natural resources** (Appendix A.3)

- nature and geological conservation site records;
- National Sand Dune Survey;
- British Geological Survey mineral assessment reports.

This basic information can be compiled into **inventory maps** of coastal geomorphology and resources, highlighting some or all of the following features: superficial materials, topography, man-made and natural coastal defences, littoral drift directions and the nature and trend of coastal changes. The results of this compilation exercise should form the basis of the subsequent assessments of flood risk (see Chapter 6), sediment budget and sensitivity for both planners and developers. Deficiencies in the existing data sources for a particular stretch of coast will highlight the need for specially commissioned data collection exercises.

Information Needs for Planners

(a) **Minerals Local Plan Policy Preparation**; it has been recognised that mineral extraction will generally be inappropriate in sensitive sand dune areas because of the potential for increasing the risk of flooding and degradation of conservation and amenity value (PGG 20; DoE, 1992). Should such areas be considered as potential mineral working sites it will be necessary to establish whether, in consultation with the NRA, local authority engineers and other responsible bodies:

- the removal of sand would have significant implications for the regional sediment budget;
- the stability of the dune systems will be significantly affected by the removal of material;
- any changes in sand dune stability will lead to increased flood risk;
- the operations would lead to significant reduction in conservation or amenity value;
- environmental protection measures could be employed to mitigate against the adverse effects of the operations.

Table 8.1 outlines the key sources of information and specialist advice that are relevant to the **general assessment** of the various considerations associated with sand dune areas. Of particular importance is the need to seek specialist advice from the NRA and relevant Coastal Defence Groups (in England and Wales) and the appropriate Regional Council department (in Scotland).

(b) **Allocation of Land for Development Plan Proposals**; In many areas sand dunes will not provide suitable sites for most forms of development due to the flood risk and the implications for the conservation and amenity value. Should sites be considered to be potential development areas, the local planning authority should consult with local authority engineers or the NRA to establish whether the site can be safely developed, taking account of the need for new or improved flood defences and the possible environmental effects (see Chapter 6)

Table 8.1 Sand dunes: Key sources of information and advice for general assessments.

Key Considerations	England and Wales	Scotland	Sources of Specialist Advice
Risks: Tidal Flooding	• NRA maps of areas prone to flooding (S.105 Maps) • NRA Sea Defence Survey	Various possible sources, including: • Island and Regional Council Water Services Depts. or equivalent • River Purification Authority • Ordnance Survey maps can be used to identify the extent of land below 5m	NRA; Scottish Regional Councils; Coastal Defence Groups; Proudman Oceanographic Laboratory; Storm Tide Warning Service of the Meteorological Office
Risks: Wind blown sand	• Historical records • Meteorological data • Soil survey maps and memoirs	• Historical records • Meteorological data • Soil survey maps and memoirs • The Beaches of Scotland	Specialist consultants.
Sediment budget	• Shoreline Management Plans • British Geological Survey maps and memoirs • Macro–Review of the Coastline (HR Wallingford)	• Shoreline Management Plans • British Geological Survey maps and memoirs • The Beaches of Scotland	NRA; Scottish Regional Councils; Coastal Defence Groups Coast protection authorities
Sensitivity	• Shoreline Management Plans • Coastal Management Plans • Conservation agency maps and records • National Sand Dune Survey • NCC Atlas of Coastal Sites Sensitive to Oil Pollution	• Shoreline Management Plans • Conservation agency maps and records • National Sand Dune Survey • NCC Atlas of Coastal Sites Sensitive to Oil Pollution	NRA; Scottish Regional Councils; Coastal Defence Groups; Conservation agencies; JNCC; Coast protection authorities

Information Needs for Developers

Owing to the level of risks and the high conservation and amenity value, sand dune areas will rarely provide suitable sites for most forms of development. If development is contemplated in these areas, the developer would need to demonstrate that the proposals have taken full account of flooding and blown sand issues and the possible environmental effects.

9 Information Needs: Coastal Cliffs

Introduction

Erosion of coastal cliffs can present a significant hazard to land use and development. Indeed, the cumulative loss of land, cliff top properties, services and infrastructure are problems that are experienced on many coastlines. However, the nature of the problems that are encountered vary according to the geological setting and the type of event which produces cliff recession. Problems can range from the effects of rapid cliff retreat on soft cliffs, as at Holderness, to slow ground movement on unstable coastal slopes, as in Ventnor, Isle of Wight, and threats to public safety from cliff falls on hard rock coasts.

Cliff recession should be viewed as a 3–stage process comprising **detachment** of material (ranging from individual soil particles to enormous coherent blocks of material in landslides), the removal of debris and **transport** by water, and the **deposition** of sediment elsewhere (Figure 9.1). The rate of removal of debris from the foreshore is of particular importance as this material provides protection against further detachment.

Although sub–aerial processes can be important locally, most erosion of cliffed coastlines is achieved by cliff falls and landsliding and is usually stimulated by wave attack and groundwater levels in the coastal slopes. Indeed, all cliffed coastlines are testimony to the cumulative efficacy of falls and landsliding, for they are in reality, the coalescent scars of innumerable individual failures. The principle recession mechanisms are falls, slides and flows, although the latter are not particularly significant around the British coast. **Falls** occur when material becomes detached from a cliff face; they occur whenever the coast is retreating, but seldom leave a lasting trace. **Slides** involve movement along a basal shear surface and can take a variety of forms: rotational, translational or compound.

The coastline in Britain is very long and well known for its varied character, rapidly changing rock type and local intensity of marine erosion, so it should come as no surprise that the major areas of coastal recession are correspondingly diverse. For a particular geological setting and set of environmental controls (climate and sea level) there will be a characteristic set of recession processes (**cliff behaviour**) giving rise to **characteristic cliff forms**. Whilst characteristic forms develop, the individual components will be evolving and the patterns of recession continually changing. Four broad characteristic forms of coastal recession can be recognised on the basis of ground–forming materials and type of failure (Figure 9.2; Jones and Lee, 1994):

(i) **cliffs developed in weak superficial deposits;** The east coast of England from North Yorkshire to Essex and parts of North Yorkshire is largely developed in thick sequences of glacial till interbedded with sands and gravels, occasionally overlying hard rock cliffs. These deposits can be rapidly eroded by the sea; for example, the entire 60km length of the undefended Holderness coastline (Humberside) has retreated at rates of 1.8m per year since 1852. On the North Yorkshire coasts the till cliffs can be prone to major dramatic landslides; the Holbeck Hall failure of June 1993 was the most recent example.

(ii) **cliffs developed in stiff clay;** Stiff clays are particularly prone to landsliding with classic examples occurring along the southern shore of the Thames estuary in Kent, where cliffs up to 40m high developed in London Clay have repeatedly failed in response to marine erosion, which results in average retreat rates of up to 2m per year.

Figure 9.1 A 3-stage cliff recession model.

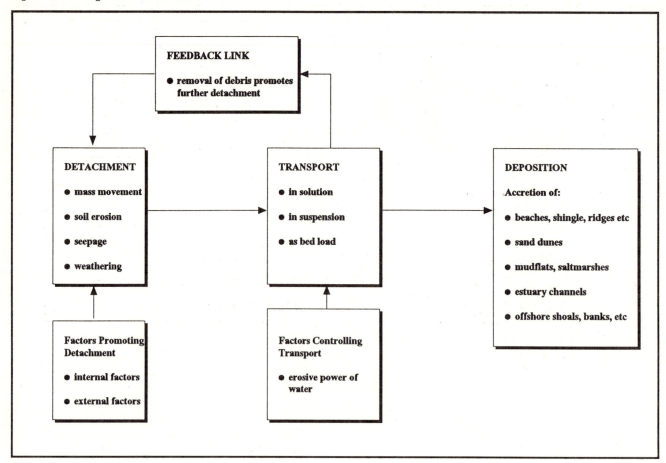

Although minor failures occur elsewhere along the coast where clay strata outcrop, the largest failures are associated with interbedded clays and sands. There are conspicuous failures of this type on the north coast of the Isle of Wight, especially at Bouldner. At Barton-on Sea in Christchurch Bay landslides extend for 5km on cliffs up to 30m high developed in Barton Clay and Barton Sand overlain by Plateau Gravel.

(iii) **cliffs developed in stiff clay with a hard cap-rock;** The largest coastal landslides occur in situations where a thick clay stratum is overlain by a rigid cap-rock of sandstone or limestone, or sandwiched between two such layers. Amongst the most dramatic examples is Folkestone Warren, Kent, where the high Chalk Cliffs have failed on the underlying Gault Clay, the Isle of Wight Undercliff and on the West Dorset coast at Black Ven.

(iv) **cliffs developed in hard rock;** Coastal cliffs developed in rocks are continually suffering minor collapse due to basal undermining by the sea. These events are most frequent in the soft rock clifflines of south-eastern England such as the famous Seven Sisters and Beach Head Chalk cliffs of East Sussex, which are currently retreating at an average rate of 0.97m a year. Large falls also occur on a number of coasts including the Triassic sandstone cliffs of Sidmouth, Devon and the Liassic limestone cliffs of South Glamorgan.

Cliff recession can be an episodic process with periods of little or no erosion separated by rapid erosion; occasionally dramatic landslides which may remove large sections of coastline in a single event. The 1993 Holbeck Hall landslide at Scarborough, for example, involved around 95m of cliff retreat over a five day period, with 60m lost overnight. The rate of recession is controlled by sequences of events which can range in frequency from 4–5 years (eg. the Holderness coast) to over 5,000 years (eg. the Isle of Wight Undercliff). Recession events should be viewed as inherently uncertain because of the complex response of different cliffs to events such as extreme rainfall or storm surges.

Figure 9.2 Characteristic eroding cliff types.

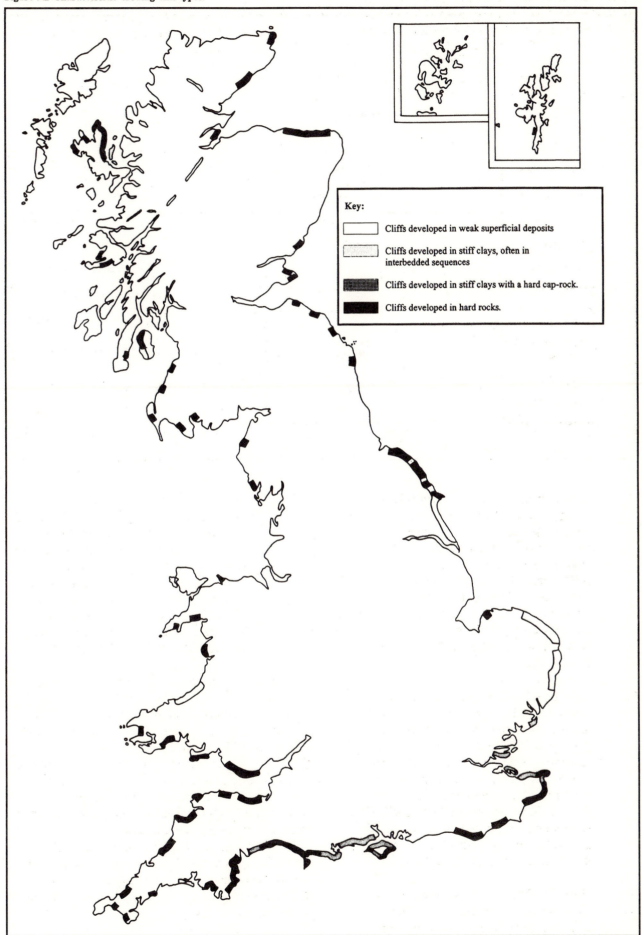

Key:

☐ Cliffs developed in weak superficial deposits

▨ Cliffs developed in stiff clays, often in interbedded sequences

▩ Cliffs developed in stiff clays with a hard cap-rock.

■ Cliffs developed in hard rocks.

The causes of cliff recession are complex and varied, most events are promoted by a combination of wave attack and groundwater levels within the cliff (Figure 9.3). The effect human activity should, however, never be underestimated, with many cliff failures arising as a result of:

- uncontrolled discharge of surface water through soakaways and highway drains;
- progressive deterioration and leakage of swimming pools and services such as foul sewers, water mains and service pipes;
- inappropriate excavations to create plots for building, especially at the foot of a cliff;
- disruption of sediment transport by groynes and breakwaters, leading to starvation of beaches and accelerated erosion.

The risk associated with cliff recession events is determined by the type of movements which can be expected to occur and their potential consequences. Although a wide variety of factors (e.g. material characteristics, geological structure, pore water pressures, slope angle, etc.) and causes (e.g. coastal erosion, weathering, seepage erosion, high groundwater levels, etc.) are important in determining the occurrence of mass movement, it is the effective **shear strength** operating at various depths within a slope and along any pre-existing shear surfaces which is critical in controlling character of failure. Three main groups of eroding cliff can be recognised (Figure 9.4):

(i) **cliffs prone to first-time failures** of previously unsheared ground, often involving the mobilization of the **peak strength** of the material. Such landslides are often characterised by large, rapid displacements, particularly if there are large differences between the peak and residual strength values. The Holbeck Hall landslide is an example of this style of failure, with the dramatic movements occurring on an intact coastal slope developed in glacial till.

Repeated failures of unsheared slopes is a common feature around the coast. Erosion of the soft glacial till cliffs of the Holderness coast, for example, involves relatively small first-time failures at a given point every 1–5 years;

(ii) **cliffs prone to failure along pre-existing lines of weakness** such as faults, joints or bedding planes. On many hard rock coasts the potential for failure and cliff recession is controlled more by the presence and pattern of these discontinuities than the strength of the rock mass itself;

(iii) **cliffs prone to reactivation of pre-existing landslides** where part or all of a previous landslide mass is involved in new movements, along shear surfaces where the materials are at **residual strength** and non-brittle. In many inland situations landslides can remain dormant or relatively inactive for thousands of years. However, in the case of coastal landslides, marine erosion removes material from the lower parts of the slopes, thereby removing passive support and promoting further movement. Such failures are generally slow moving, although more dramatic failures can occur (Hutchinson, 1987).

The importance of this distinction between first-time and pre-existing slides is that once a slide has occurred it can be made to move under conditions that the slope, prior to failure, could have resisted.

Construction of sea walls and other cliff foot structures has generally reduced the rate of recession and the likelihood of slope instability problems. However, the prevention of marine erosion does not eliminate the potential for slope failure, highlighting the importance of internal factors (such as ground water and weathering) in promoting instability. Whilst slope degradation behind defences generally involves relatively small events, large-scale dramatic events do occur and can result in considerable loss of land, including:

- 1993 Holbeck Hall landslide, Scarborough which led to the destruction of the hotel and sea walls below with a loss of around 100m of land;

- the landslide at Overstrand, Norfolk where around 100m of cliff top land was lost during a three year period between 1991 – 1994.

Both these events were **first-time failures** of protected, intact coastal slopes. Continued slope

Figure 9.3 Causes of cliff recession.

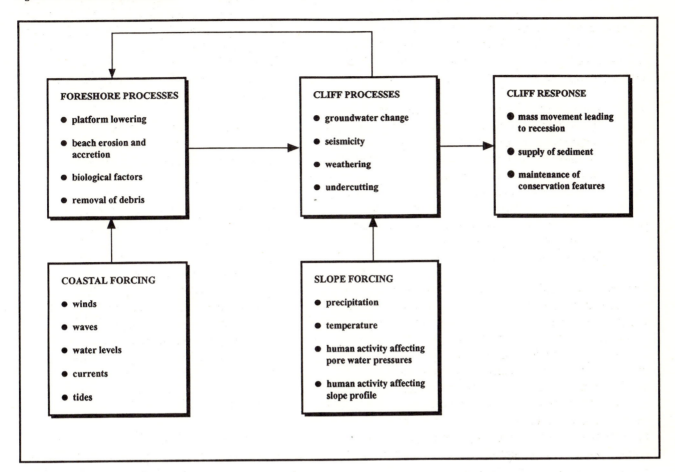

instability problems can also be experienced where cliff foot structures have been used to protect **pre-existing landslides**. For example:

- the major landslides at Barton–on Sea during 1974 (Clark et al, 1976);

- the continued ground movement problems at Sandgate (Palmer 1991), the Isle of Wight Undercliff (e.g. Rendel Geotechnics, 1995c) and the Isle of Portland (McLaren, 1983).

From these examples, it is clear that whilst protection against marine erosion can considerably enhance slope conditions it is not a panacea for preventing cliff recession, because of the complexity of many instability problems. Indeed, on cliffs affected by the first–time or repeated failures, prevention of marine erosion can result in **free degradation** where the slope angle is too steep for the materials and ground water conditions. Indeed, it is important to note that many protected cliffs may stand for a period of time before suffering failure. The mechanisms involved are probably analogous to those

experienced in the widely studied delayed failures (up to 100 years after excavation) in London Clay cuttings (e.g. Skempton, 1964; Chandler, 1984). On pre–existing slides, prevention of marine erosion alone can prevent further deterioration of stability conditions but may leave the slope as vulnerable to triggering factors, such as rainfall events.

It is also important to stress that **slope stabilisation** works (e.g. drainage or regrading) cannot generally achieve significant improvements to stability unless marine erosion is prevented. It is clear, therefore, that effective management of coastal cliffs should take account of the effects of **cliff top, coastal slope** and **shoreline conditions** on the overall stability of particular coastal sections. This type of coordinated approach has not been common practice around the coast, although the scale and complexity of problems in some areas (e.g. the Isle of Wight Undercliff, Canterbury City Council's London Clay Cliffs and the Scarborough urban area) has led to the adoption of integrated cliff management techniques.

Cliff recession can have an important role in supplying littoral sediment to beaches, sand dunes and mudflats on adjacent stretches of coastline. The

Figure 9.4 Geotechnical classification of eroding coastal cliffs.

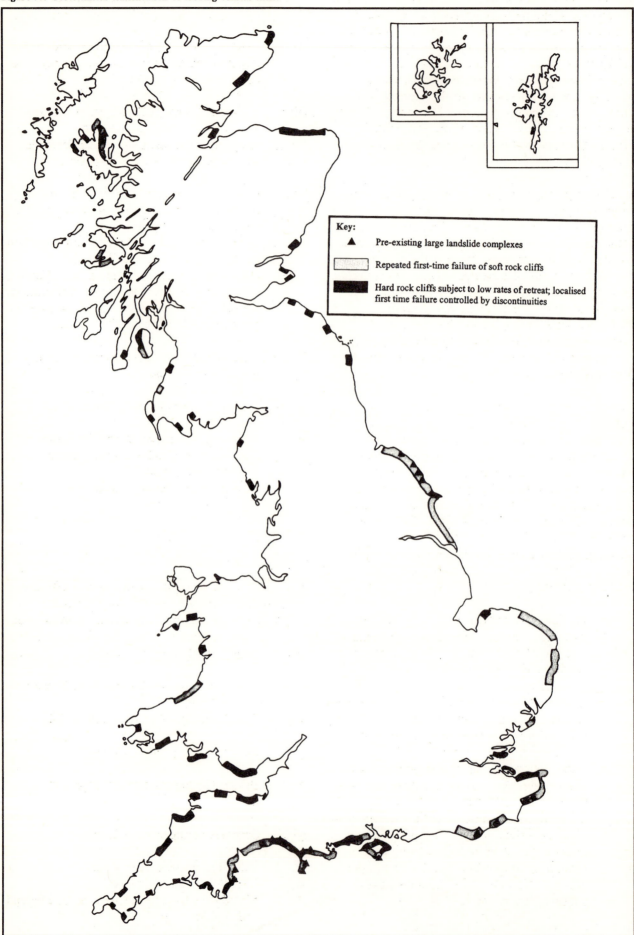

Key:

▲ Pre-existing large landslide complexes

Repeated first-time failure of soft rock cliffs

Hard rock cliffs subject to low rates of retreat; localised first time failure controlled by discontinuities

annual input of sediment into the coastal zone from the eroding Holderness coastline, for example, is believed to be around $1Mm^3$ of which $0.7Mm^3$ is fine grained material, some of which is carried into the Humber estuary. The coarser material ($0.3 \, Mm^3$ a year) is moved southwards to Spurn Head and, possibly, across the Humber to the beaches and dunes of the Lincolnshire coast. Disruption of this sediment supply, by the use of erosion control techniques, may lead to an increase in erosion or flood risk elsewhere.

Coastal cliffs are frequently of great conservation value as breeding grounds for seabirds or by supporting important habitats. Cliffs are also of great value because of the exposures of geological features such as important fossil beds or marker horizons for rocks of a particular age. Continued erosion is necessary to ensure that these features remain accessible for teaching and research purposes. Some coastal landslides are also of particular importance as teaching and research sites for mass movement studies and, consequently, have been designated as SSSIs.

Coastal landsliding and cliff recession are important considerations for both planners and developers. Indeed, development in unsuitable locations can lead to a range of problems from adverse effects on the stability of adjacent land to calls for publicly funded protection measures and the consequent effects on conservation or coastal defence interests elsewhere. As both PPG 14 and PPG 20 (DoE, 1990, 1992a) have recognised, the planning system clearly has an important role in minimising the risks associated with coastal instability through:

- guiding development away from unsuitable locations. This may involve establishing "set–back" lines within which development could be affected by erosion over a specified period;
- ensuring that development does not initiate or exacerbate instability problems on adjacent land, by specifying appropriate site drainage requirements and limiting slope excavation during development etc.;
- ensuring that the precautions that are taken to minimise risks from cliff instability do not lead to starvation of sediment supply to other important coastal sites and, thereby, increase the level of risk elsewhere;
- ensuring that development does not have adverse effects on local amenities and conservation interests.

The key considerations for cliffed coastlines are thus:

(i) **risks**; the impact of landsliding and cliff recession on development and the need for coast protection works;

(ii) **sediment budget**; the effects of erosion control at a particular site on the sustainability of natural coastal defences and recreational beaches elsewhere;

(iii) **sensitivity**; the effects of erosion control measures on the integrity of geological conservation sites; the effects of development and coast protection works on the level of erosion and flood risk elsewhere in a coastal system; the effects of sea level rise on the rate of cliff erosion; the effect of changes in beach management practices on the rate of cliff erosion.

These considerations have been examined in the following trial study areas:

- Seaham – Teesmouth, Cleveland and Durham;
- Filey – Scarborough, North Yorkshire;
- Flamborough Head – Outer Humber Estuary, Humberside;
- Cromer – Sea Palling, Norfolk.

Background Information

Background sources relevant to coastal cliffs can be identified from the Coastal Zone Database (Appendix A.1). The most important sources are:

(i) **landforms** (Appendix A.2.1):

- The NCC Atlas of Coastal Sites Sensitive to Oil Pollution;
- Aerial photographs (Appendix A.5.4).

(ii) **materials** (Appendix A.2.2):

- British Geological Survey maps and memoirs.

(iii) **processes** (Appendix A.2.3):

- Shoreline management plans.
- UK Digital Marine Atlas;
- Macro–Review of the Coastline of England and Wales;

- The Beaches of Scotland;
- Review of Landsliding in Great Britain;
- Review of Erosion, Deposition and Flooding in Great Britain;

This basic information can be compiled into an **inventory map** of coastal geomorphology, highlighting some or all of the following features: bedrock geology, superficial materials, landslide features, topography, slope steepness, man–made and natural coastal defences, littoral drift directions and the nature and trend of coastal change. The results of this compilation exercise should form the basis of the subsequent assessments of risks, sediment budget and sensitivity for both planners and developers. Deficiencies in the existing data sources for a particular stretch of coast will highlight the need for specially commissioned data collection exercises.

Information Needs for Planners

(a) **Strategic and Local Plan Policy Preparation**; Table 9.1 outlines the key sources of information and specialist advice that are relevant to the **general assessment** of coastal cliff conditions. Here, defining areas where particular consideration should be given to the effects of cliff erosion involves identifying the **distribution of coastal landslides** and **predicting cliff recession rates**, as outlined below:

(i)   **coastal landsliding**; the National Review of Landsliding in Great Britain (inland as well as coastal; Appendix A.2.3), is a census of reported landslides which revealed grand total of 8835 landslide features (Figure 9.5); 6120 were recorded in England of which 1013 are coastal, 1200 in Scotland (175 on the coast) and 1515 in Wales (114 on the coast). Distribution maps of landslides recorded in the **published literature** have been produced for all counties in England and Wales and for Scotland. These maps are available at a scale of 1:250,000 and show the distribution of landslides recorded in the original census (completed in 1985). Each recorded landslide is portrayed with a unique reference number and is described in terms of landslide type and age. In addition, areas of suspected but unrecorded instability are also shown (Figure 9.6).

To encourage the wide use of the survey results the information has been compiled in a computerised archive– the **National Landslide Databank**. This database includes the results of the original census and also landslides identified in an updating programme that was completed in 1991. Landslide information for an individual county or Scottish region can be purchased on computer discs for direct access using an IBM PC or fully compatible desk top computer. Alternatively, a telephone enquiry service is available for identifying recorded landslides within a specified area. The charge covers the cost of providing location data, a database listing for each landslide record, and relevant bibliographic references.

The National Landslide Databank and associated maps can provide a general indication of where coastal landslide problems are likely to occur. Further information and the nature and cause of potential problems can be obtained by accessing the original source material highlighted by the bibliography attached to the database.

Although the Databank represents the most complete data set on the nature, occurrence and extent of landsliding in Great Britain, it is wholly composed of **reported occurrences** and is, therefore, an artifact of knowledge rather than a true representation of the actual distribution of coastal cliff erosion problems. It must be anticipated that in many areas the information that can be obtained from the Databank will need to be supplemented by other sources to gain a realistic appreciation of the **actual pattern** of landsliding. Probably the most cost-effective information source are vertical aerial photographs available from the Ordnance Survey or commercial air survey firms (see Appendix A.5.4). Coastal landslides can be readily identified from aerial photographs because of their distinctive surface form and tonal patterns (Table 9.2). The key to successful interpretation is a systematic search of the coastal cliffs where instability may be anticipated. The most appropriate photograph scale is probably around 1:25,000; although larger scales provide more detail they will require considerably

Table 9.1 Coastal cliffs: Key sources of information and advice for general assessments.

Key Considerations	Key Sources of Information	Sources of Specialist Advice
Risks: Landsliding and Cliff Recession	● Shoreline Management Plans ● National Landslide Databank ● Coast Protection Surveys (England and Wales) ● British Geological Survey maps and memoirs ● Ordnance Survey maps ● Aerial photographs	Coast protection authorities; Coastal Defence Groups Geotechnical consultants
Sediment Budget	● Shoreline Management Plans ● British Geological Survey maps and memoirs ● Macro–Review of the Coastline	Coast protection authorities; Coastal Defence Groups
Sensitivity	● Shoreline Management Plans ● Conservation agency maps and records ● NCC Atlas of Coastal Sites Sensitive to Oil Pollution ● Ordnance Survey maps	Coast protection authorities; Coastal Defence Groups; Conservation agencies

more effort to inspect large lengths of coast

(ii) **prediction of cliff recession rates**; the most readily accessible sources of information about coastal change are Ordnance Survey maps and historical maps (Appendix A.5.1). Charting cliff retreat between maps of different dates might appear straightforward, but in fact such sources can frequently be misleading. Early maps are often unreliable in the depiction of cliffs in undeveloped areas, the accurate representation of relief and landshape in areas of rapid erosion or active landslide complexes would not have been a high priority for the early Ordnance Survey surveyors and their predecessors. Indeed, the depiction of steep slopes by means of hachuring often owes more to artistic licence than to an accurate representation of topographic form. In some instances, there may be significant differences in **map projection** between sets of historical maps.

Analysis of historical maps can give an indication of the **average annual rate of erosion** along a particular stretch of coast. However, cliffs do not retreat in a uniform manner; long periods of relative inactivity are separated by infrequent large events. For example, during the 1953 east coast storm surge, glacial sand cliffs at Covehithe retreated over 13m overnight, in comparison with long–term average rates

Figure 9.5 National Review of Landsliding: Summary distribution map (after Jones and Lee, 1994).

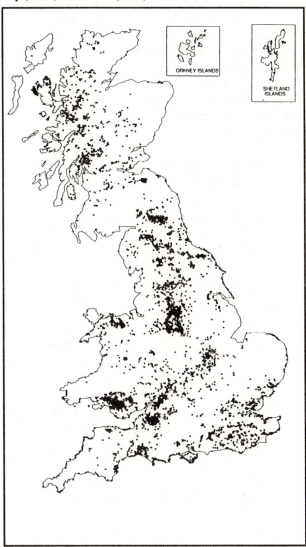

Figure 9.6 National Review of Landsliding: Sample of a 1:250,000 scale landslide distribution map.

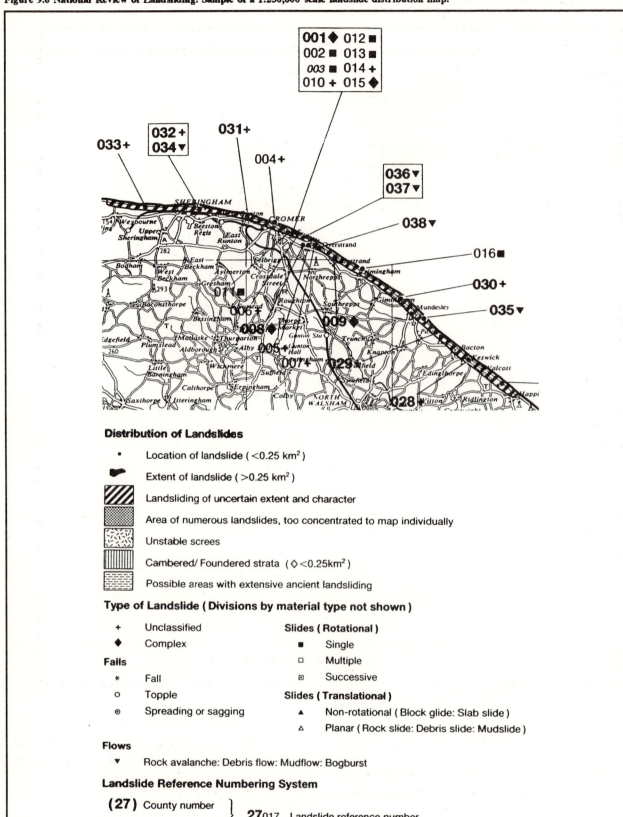

Distribution of Landslides

- • Location of landslide (<0.25 km²)

- ◣ Extent of landslide (>0.25 km²)

- ▨ Landsliding of uncertain extent and character

- ▦ Area of numerous landslides, too concentrated to map individually

- ▨ Unstable screes

- ▥ Cambered/ Foundered strata (◇ <0.25km²)

- ▤ Possible areas with extensive ancient landsliding

Type of Landslide (Divisions by material type not shown)

+	Unclassified	**Slides (Rotational)**
♦	Complex	■ Single
Falls		□ Multiple
*	Fall	⊡ Successive
o	Topple	**Slides (Translational)**
⊙	Spreading or sagging	▲ Non-rotational (Block glide: Slab slide)
		△ Planar (Rock slide: Debris slide: Mudslide)

Flows

▼   Rock avalanche: Debris flow: Mudflow: Bogburst

Landslide Reference Numbering System

(27) County number
017   Slide number
} **27**017   Landslide reference number

Landslide Age (represented by size and style of number)

017   Active (currently unstable or cyclically active with period up to 5 years)

017   Recent (movement within last 100 years)

Table 9.2 Typical indicators of slope instability from aerial photographs.

PROCESS	KEYS TO RECOGNITION ON AERIAL PHOTOGRAPHS
Rockfall	• light striations of freshly exposed rock
	• stripped vegetation
	• cones of light coloured fresh scree
Rotational failure	• arcuate backscar above concave slump feature
	• intact strata exposed in failure scar
	• depression behind back–tilted strata, possibly occupied by a pond or lake
	• oval or lens shaped areas of cultivation
	• walls destroyed, trees curved at base or leaning back
	• rivers displaced abruptly
	• cliff forms change suddenly
Translational or planar side	• geological outcrops dipping in same direction as ground
	• ruckled or uneven ground in front of step feature
	• angular backscars or crevices following joints
Debris slide	• light coloured areas with flow patterns
	• road re–alignments
Rotational slump	• arcuate backscars above concave slump feature
	• dark tones over damp or waterlogged ground
	• lobate toe at advancing edge of slump
Debris and mud flow	• hummocky, stepped or fissured ground avoided by cultivation
	• displaced boundaries, disrupted vegetation patterns
	• patterns of light and dark tones
	• isolated boulders remote from outcrop
	• debris outwash fans on valley side
	• cones of loose, unvegetated scree

of around 5m per year.

Vertical aerial photographs provide a more reliable source of information about cliff erosion rates, although they are generally only available from around 1940 onwards. In contrast to many maps, photographs allow clear identification of the evolution of coastal landslide systems as well as the accurate determination of the pattern of cliff retreat. They do, however, suffer from the same limitations as maps in that they are incidental observations made at a fixed time and, hence, tend to only give limited indications of the nature and rate of contemporary processes.

Measurement of recent and current erosion rates can be used to define an area which is likely to be affected by instability or erosion in future decades. These estimates can be improved through an appreciation of the nature and scale of individual landslide events that could be expected on different stretches of cliff. It should be appreciated, however, that predictions of coastal changes will form part of shoreline management plans currently being prepared by coastal defence groups. In order to prevent unnecessary duplication of effort, local planning authorities should make full use of this information. In this context, it is worth noting that MAFF have recently established a programme of research to develop methods of predicting soft cliff recession rates suitable for both detailed and strategic planning. (Rendel Geotechnics, in prep.).

(b) **Allocation of Land for Development Plan Proposals**; Local planning authorities should establish whether a shortlisted site can be safely developed without the need for improved slope stabilisation or coast protection works. Should new preventative measures be required the authority should consider:

- whether the necessary slope stabilisation or coast protection works can be provided without adversely affecting other interests, including: conservation and coastal defence interests elsewhere;
- the possible cost implications of the new or improved defences to potential developers and whether they are likely to be economical;
- the possible nature and scale of new or improved defences that would, in principle, be acceptable.

This should involve undertaking a **site review** of the shortlisted or potential sites. In most instances, this review would involve no more than seeking advice from local authority coastal engineers about predicted recession rates. However, where instability issues remain to be resolved after consultation it may be necessary to carry out specific site studies, comprising:

(i) **ground inspection;** aerial photograph interpretation and surface mapping techniques such as geomorphological mapping can be used to establish the nature and extent of possible pre–existing landslides in an area and to determine the degree of threat that these hazards may pose to existing property and future developments. Geomorphological mapping was carried out, for example, as part of an investigation to establish the cause of the 1987–88 landsliding at Luccombe, Isle of

Wight which had resulted in a number of houses being damaged beyond repair (Lee & Moore 1989). By carefully mapping the limits of the recent movements, together with the surface form of the surrounding area, it was possible to demonstrate that the village had been built on an ancient landslide and that the recent movements were a reactivation of parts of this feature (Figure 5.6). It must be recognised, however, that such qualitative assessments can be fallible. Slopes may be so remodelled by human activity or natural processes that no surface indicators exist to indicate the presence of failed masses and buried shear surfaces.

(ii) **assessment of landslide risk;** assessment of landslide risk involves the evaluation of the key factors which influence the pattern and character of landslide activity and considering the potential consequences of slope failure. The variable interaction amongst specific aspects of these factors helps to determine the precise siting of instability, the timing of failure and to dictate the condition of displacement in terms of the form, rate and duration of movement. However, there are a broad range of causal factors that may be important; some trigger instability through short term fluctuations, while others act in an insidious manner to slowly reduce the resistance of a cliff to failure.

The great variety of landslide forms and causes means that landslide risk assessment techniques have to be adapted to the local conditions. However, a number of principles are likely to apply in most cases:

- establish the type and size of failure that could occur at the shortlisted sites. This should involve identifying the nature and extent of past and present failures along the entire outcrop of the materials that underlay the sites;
- establish a historical record of events, their size and the impact of property, services and infrastructure;
- establish a relationship between historical events and rainfall patterns or storm tide levels. This will enable a broad assessment to be made of the likelihood of an event occurring somewhere on the outcrop of the site materials as a result of a particular rainfall period or wave height;

- establish what areas could be affected by further landsliding and over what time period.

These principles are well illustrated by the DoE – commissioned assessment of Coastal Landslip Potential for Ventnor, Isle of Wight (Lee & Moore, 1991). Ventnor is an unusual situation in that the whole town lies within an ancient landslide complex. The problems are related to the control of the nature of development in those parts of the town which have been shown to be particularly susceptible to ground movement. The hazards faced by the local community in Ventnor have been defined in terms of an understanding of **contemporary ground behaviour** within an extensive belt of landslipped ground.

The approach developed for assessing the hazards at Ventnor involved a thorough review of available records, reports and documents relating to instability, followed by a detailed field investigation comprising geomorphological and geological mapping, photogrammetric analyses, a survey of damage caused by ground movement, a land use survey and a review of local building practice (Figure 9.7). The results of these investigations provided an understanding of the nature and extent of the landslide system, together with the type, size and frequency of contemporary movements and their impact on the local community. This detailed understanding of ground behaviour was used, in conjunction with knowledge of the vulnerability to movement of different types of construction and the spatial distribution of property at risk, to formulate a range of management strategies designed to reduce the impact of future movements.

Geomorphological mapping revealed the scale and complexity of the landslides; once the framework of landslide units had been established, it was possible to relate building damage and movement rates to units with known dimensions. A search through historical documents, local newspapers from 1855–1989, local authority records and published scientific research, revealed nearly 200 individual incidents of ground movement over the last two centuries. The landslide–related information was presented as a 1:2500 scale map of **ground behaviour** which summarised the nature, magnitude and frequency of contemporary processes and their impact on the local community. The behaviour map demonstrated that the problems resulting from ground movement vary from place to place according to the geomorphological setting. This formed the basis for **landslide management strategies** that can be applied within the context of a zoning framework that reflects the variations in stability rather than a blanket approach to the problem. In support of the management strategy a 1:2500 scale Planning Guidance map was produced which related categories of ground behaviour to forward planning and development control (Figure 9.8, Table 9.3). The map indicated that different areas of the landslide complex need to be treated in different ways for both policy formulation and the review of planning applications. Areas were recognised which are likely to be suitable for developments, along with areas which are either subject to significant constraints or mostly unsuitable.

A similar procedure was used to provide a preliminary assessment of the risks associated with potential cliff instability in Scarborough's South and North Bays, following the 1993 Holbeck Hall failure (Rendel Geotechnics, 1994). Potential problems vary considerably along this coast, according to the geomorphological setting and the type of failure that may occur. On this basis a general indication of the level of risk was achieved by defining:

- **landslide systems** characterised by a unique range of instability problems (ie. **contemporary processes**; (Figure 9.9);

- the **landslip potential** within each landslide system, expressed in terms of the likelihood of different types of failure that may occur (Figure 9.9);

- the **elements at risk** from instability ie. property, infrastructure and services on and adjacent to individual landslide systems (Table 9.4)

- the severity of the potential consequences of each possible type of landslide event within each system (**potential impact**); the consequences were considered on a scale of losses from **minor** (inconvenience and slight to moderate damage), **partial** (moderate to severe property damage) to **total** (injury and severe damage to complete destruction Table 9.4).

(iii) **preliminary assessment of environmental effects** of possible development and slope stabilisation and coast protection scheme options, in consultation with the coast protection authority, conservation agencies and other interested groups.

Figure 9.7 The Ventnor Landslip Potential Study: Programme of work (after Lee and Moore, 1991).

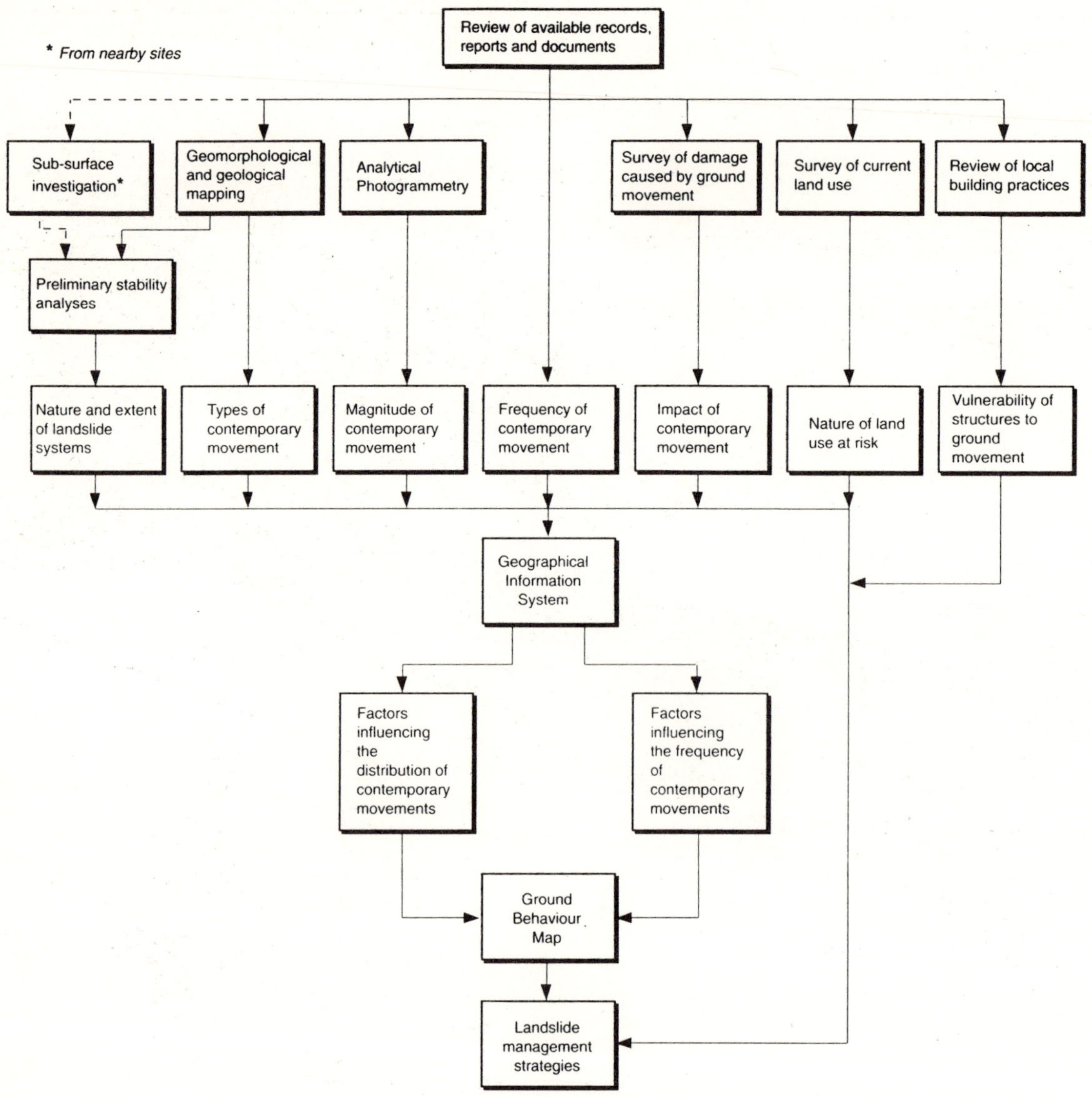

```
                                    ┌──────────────────────────┐
                                    │ Review of available records,│
                                    │ reports and documents     │
                                    └──────────────────────────┘
    * From nearby sites
┌──────────────┐ ┌──────────────┐ ┌──────────────┐  ┌──────────────┐ ┌──────────────┐ ┌──────────────┐
│ Sub-surface  │ │Geomorphological│ │ Analytical   │  │ Survey of damage│ │ Survey of current│ │ Review of local│
│ investigation*│ │and geological │ │ Photogrammetry│  │ caused by ground│ │ land use     │ │ building practices│
│              │ │ mapping       │ │              │  │ movement      │ │              │ │              │
└──────────────┘ └──────────────┘ └──────────────┘  └──────────────┘ └──────────────┘ └──────────────┘
┌──────────────┐
│ Preliminary stability│
│ analyses     │
└──────────────┘
┌──────────────┐ ┌──────────────┐ ┌──────────────┐ ┌──────────────┐ ┌──────────────┐ ┌──────────────┐ ┌──────────────┐
│ Nature and extent│ │ Types of    │ │ Magnitude of│ │ Frequency of│ │ Impact of   │ │ Nature of land│ │ Vulnerability of│
│ of landslide │ │ contemporary│ │ contemporary│ │ contemporary│ │ contemporary│ │ use at risk │ │ structures to│
│ systems      │ │ movement    │ │ movement    │ │ movement    │ │ movement    │ │             │ │ ground movement│
└──────────────┘ └──────────────┘ └──────────────┘ └──────────────┘ └──────────────┘ └──────────────┘ └──────────────┘
                                    ┌──────────────┐
                                    │ Geographical │
                                    │ Information   │
                                    │ System        │
                                    └──────────────┘
        ┌──────────────┐                        ┌──────────────┐
        │ Factors      │                        │ Factors      │
        │ influencing  │                        │ influencing  │
        │ the          │                        │ the frequency│
        │ distribution of│                      │ of           │
        │ contemporary │                        │ contemporary │
        │ movements    │                        │ movements    │
        └──────────────┘                        └──────────────┘
                        ┌──────────────┐
                        │ Ground       │
                        │ Behaviour    │
                        │ Map          │
                        └──────────────┘
                        ┌──────────────┐
                        │ Landslide    │
                        │ management   │
                        │ strategies   │
                        └──────────────┘
```

Information Needs for Developers

(a) Selection of Sites for Development; in seeking to decide where development might be most suitably located, developers will need to be forewarned of the potential problems so that they can estimate the cost implications. **Site reconnaissance** (preliminary assessment) surveys should be undertaken to establish whether:

- the site could be at risk from landsliding or cliff recession during the lifetime of the development;
- there is a need for new or improved slope stabilisation or coast protection works;
- the necessary preventative measures can be provided without affecting local amenities

and conservation interest.

Much of the necessary information may be collated from previously collected data, especially the results of **general assessments** or **shoreline management studies,** if they are publicly available. These sources may provide ready access to information that can be relevant to some or all of the developer's shortlisted sites. However, as these are essentially broad−brush studies, they will need to be supplemented by specific information about conditions at the individual sites, including:

(i)   ground inspection (see above);

(ii)   preliminary assessment of landslide and erosion risk, including a review of the

Figure 9.8 Summary planning guidance map, Ventnor, Isle of Wight (See Table 9.3; after Lee and Moore, 1991).

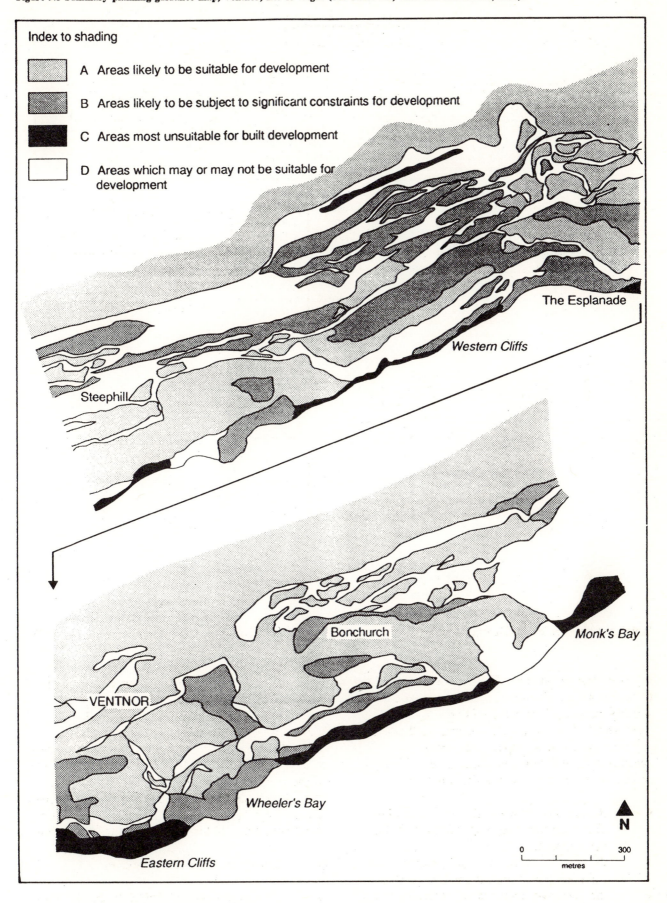

Index to shading

A Areas likely to be suitable for development

B Areas likely to be subject to significant constraints for development

C Areas most unsuitable for built development

D Areas which may or may not be suitable for development

The Esplanade

Western Cliffs

Steephill

Bonchurch

Monk's Bay

VENTNOR

Wheeler's Bay

Eastern Cliffs

N

0   300
     metres

Table 9.3 Planning guidance categories for management of landslide problems in Ventnor, Isle of Wight (after Lee and Moore, 1991).

CATEGORY	DEVELOPMENT PLAN	DEVELOPMENT CONTROL
A	Areas likely to be suitable for development. Contemporary ground behaviour does not impose significant constraints on Local Plan Development proposals.	Results of a desk study and walkover survey presented with all planning applications. Detailed site investigations may be needed prior to planning decision if recommended by the preliminary survey.
B	Area likely to be subject to significant constraints on development. Local Plan development proposals should identify and take account of the ground behaviour constraints.	A desk study and walkover survey will normally need to be followed by a site investigation or geotechnical appraisal prior to lodging a planning application.
C	Area most suitable for built development. Local Plan development proposals subject to major constraints.	Should development be considered it will need to be preceded by a detailed site investigation geotechnical appraisal and/or monitoring prior to any planning applications. It is likely that many planning applications in these areas may have to be refused on the basis of ground stability.
D	Areas which may or may nor be suitable for development but investigations and monitoring may be required before any Local Plan proposals are made.	Areas need to be investigated and monitored to determine ground behaviour. Development should be avoided unless adequate evidence of stability is presented.

relative risks at the shortlisted sites (see above) and the potential effects of development on adjacent slopes or neighbouring coastlines;.

(iii)   preliminary assessment of environmental effects of possible development, slope stabilisation, and coast protection scheme options at the different sites.

On the basis of the information collected and compiled during a site reconnaissance survey it should be possible to determine the feasibility of developing a particular site, taking account of the costs of any necessary preventative works and environmental protection measures.

(b) Detailed Design for Development; should the site reconnaissance indicate that the selected site is likely to be affected by instability or cliff erosion, then **detailed investigations** should be undertaken to:

- demonstrate to the local planning authority that the proposed development takes full account of instability or cliff erosion issues;
- assess the level of risk and determine the technical, economic and environmental feasibility of the proposed preventative work options;
- establish the design criteria for preventative works.

Successful design of developments and defence works will require a thorough investigation of the ground conditions and physical processes operating at the selected site and in the surrounding area. This could typically involve a combination of desk study and ground investigation comprising: sub-surface investigation, surface monitoring, hydrological monitoring, laboratory testing and stability analyses (Rendel Geotechnics, 1995d). The scale of any ground investigation should be discussed beforehand with the local authority and will depend on the nature of the problem at a particular site. However, it is likely that investigations will need to consider:

- the foundation conditions at the site;
- the degree of **risk** from landsliding and cliff recession;
- the nature and scale of preventative works required to reduce the risks to an acceptable level;
- the possible range of schemes that could overcome the problems, identifying suitable sources of materials and their costs;
- the likely effects of the proposed works on the local and regional **sediment budgets**;
- the **sensitivity** of the coastline at and around the site, and the likely effects of development on coastal features;
- the mitigation measures that might be used to reduce any undesirable effects of the works.

Figure 9.9 Summary of coastal instability risk, South Bay, Scarborough (after Rendel Geotechnics, 1994).

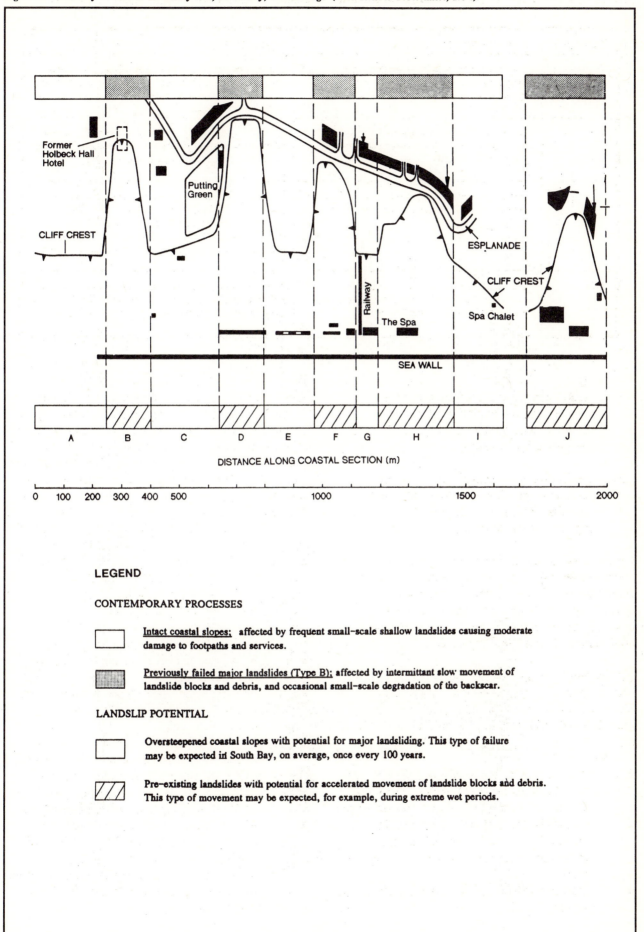

LEGEND

CONTEMPORARY PROCESSES

Intact coastal slopes: affected by frequent small-scale shallow landslides causing moderate damage to footpaths and services.

Previously failed major landslides (Type B): affected by intermittant slow movement of landslide blocks and debris, and occasional small-scale degradation of the backscar.

LANDSLIP POTENTIAL

Oversteepened coastal slopes with potential for major landsliding. This type of failure may be expected in South Bay, on average, once every 100 years.

Pre-existing landslides with potential for accelerated movement of landslide blocks and debris. This type of movement may be expected, for example, during extreme wet periods.

Table 9.4 A summary of the risk from coastal instability: South Bay, Scarborough (after Rendel Geotechnics, 1994).

Section	Setting	Elements at risk	Likelihood of failure	Potential impacts
A	Intact unprotected coastal slope with till cover thinning to south.	• Cliff top car park • Properties on cliff road • Footpaths & Cleveland Way	Small–scale failures of coastal slope could occur most years.	Slight to moderate damage to footpaths.
			Cliff recession as a result of marine erosion can be expected at an average rate of around 0.5m per year.	Loss of land, ultimately leading to re-location of Cleveland Way.
B	Previously failed (1993) major landslide.	Coast protection and slope stabilisation works have significantly reduced the potential for cliff top recession and, hence the risk to surrounding land and property.		
C	Intact protected coastal slope.	• Cliff footpaths • Putting green • Property on cliff road • Seawall	Small–scale shallow failures of coastal slope could occur most years.	Slight to moderate damage to footpaths.
			Large failures could occur involving rapid cliff top recession and runout of debris. Such a failure can be expected in South Bay, on average once every 100 years.	Loss of cliff top property and putting green, loss of coastal defences.
D	Previously failed (1885) major landslide.	• Footpaths • Shelters on lower slopes • Esplanade behind rear scarp • Seawall	Small–scale shallow failures could occur most years.	Slight to moderate damage to footpaths and shelters.
			Slow intermittent ground movement.	Slight to moderate damage to footpaths, shelters and seawalls.
			Cliff top recession as a result of degradation of the rearscarp may involve occasional small-scale failures.	Moderate to serious damage to Esplanade.
E	Intact protected coastal slope.	• Footpaths • Shelters on lower slopes • Esplanade • Seawall	Small–scale shallow failures of coastal slope could occur most years.	Slight to moderate damage to footpaths and shelters caused by ground movement and runout of debris.
			Large failure could occur involving rapid cliff top recession and runout of debris. Such a failure can be expected, on average, once every 100 years.	Loss of cliff top gardens and parts of the Esplanade, loss of coastal defences
F	Previously failed (unknown date) major landslide.	• Footpaths • Shelters • Esplanade and cliff top property • Seawall	Small–scale failures could occur most years.	Slight to moderate damage to footpaths and shelters.
			Slow intermittent ground movement could occur most years.	Slight to moderate damage to footpaths, shelters and sea walls.
			Cliff top recession as a result of degradation of the rearscarp may involve occasional small scale failures.	Moderate to serious damage to Esplanade.

Table 9.4 (cont ...)

Section	Setting	Elements at risk	Likelihood of failure	Potential impacts
G	Intact protected coastal slope.	• Footpaths • Shelters on lower slopes • Esplanade • Seawall	Small-scale failures of coastal slope could occur most years.	Slight to moderate damage to footpaths, cliff railway, buildings, and shelters.
			Large failures could occur involving rapid cliff top recession and runout of debris. Such a failure can be expected, on average, once every 100 years.	Loss of cliff property: cliff railway, footpaths and shelters.
H	Previously failed (1737) major landslide.	• Footpaths • Esplanade and cliff top property • Seawalls	Small-scale shallow failures could occur most years.	Slight to moderate damage to footpaths etc.
			Slow intermittent ground movement could occur most wet years.	Slight to moderate and serious damage to buildings, footpaths and sea walls.
			Cliff top recession as a result of degradation of the rearscarp may involve occasional small-scale failures.	Moderate to serious damage to the Esplanade.
I	Intact protected coastal slope.	• Footpaths • Esplanade and cliff top property • Seawalls	Small-scale shallow failures of coastal slope could occur most years.	Slight to moderate damage to footpaths.
			Large failures could occur involving rapid cliff top recession and runout of debris. Such a failure can be expected, on average, once every 100 years.	Loss of buildings, cliff top gardens and parts of the Esplanade, loss of coastal defences.
J	Previously failed (unknown date) major landslide.	• Cliff property • Tramways • Gardens and shelters	Small-scale shallow failures could occur most years.	Slight to moderate damage to property and retaining walls.
			Slow intermittent ground movement could occur most wet years.	Slight to moderate damage to property and retaining walls.
			Slight cliff top recession may occur.	Slight to moderate damage to cliff top property.

10 Management of Earth Science Information

Introduction

Effective management of the earth science information collected and compiled by local authorities and other organisations should be viewed as essential for ensuring its efficient use by both planners and developers. Traditional forms of information management have involved indexed archives of reports and maps (**manual systems**). Although such approaches are unlikely to become redundant, it is becoming apparent that computer-based systems have considerable potential for enhancing:

- the storage and retrieval of information;

- the accessibility of information to decision-makers;

- the updating of reports and interpretative maps and the revision of planning guidance;

- the ability to perform rapid scientific analyses on the information.

There have been considerable advances in computer technology in recent years coupled to significant reductions in hardware and software costs. This has been coupled to greater ease of use of computer technology leading to greater accessibility to IT methods by an increasing number of personnel. However, it should not be assumed that a computerised solution will always be the most cost-effective route for local authorities. The key to successful information system development lies in gearing the development to;

- the perceived users of the system (in terms of background and expertise),

- the information and knowledge that those users wish to extract from the system,

- the complexity and relationships between the information in the study area,

- the resources available for development (hardware, personnel, cost),

- the time constraints of development.

Appropriate information systems are those which achieve the desired objectives within the constraints. By carrying out a detailed **requirements analysis** it is possible to identify objectives and constraints and to develop a suitable management approach. These will lie somewhere along a spectrum ranging from totally manual systems at one end to complex geographic systems at the other. Systems developed within this spectrum can generally be classified into one of the following classes:

- totally manual systems,
- simple databases,
- geo-referenced databases,
- geographic information systems – ranging from simple single seat installations through to complex multi-seat, multi-site operations.

The relative attributes of each particular type of system are summarised in Table 10.1. This Chapter examines the development and use of information management, highlighting the relative attributes of a range of systems of different levels of sophistication. The role of geographic information systems (GIS) is examined with reference to a demonstration system developed as part of this study (the Ventnor GIS; see Annex B).

Table 10.1 Relative suitability of different information system types.

	Manual System	Simple Database	Geo-referenced Database	Geographic Information System
Cost	Very low	Low	Low–Moderate	Moderate–High
Implementation time	Low	Low–Moderate	Low–Moderate	Moderate–High
Maintenance cost	n/a	Low	Low–Moderate	Moderate–High
Ease of use	Poor	Good	Good	Fair
Functionality	Very Poor	Good	Good	Good
Analytical capability	Very Poor	Fair	Good	Very Good
Tabular Output	No	Yes	Yes	Yes
Automated Map Output	No	No	Yes	Yes
Use accessibility to information	Very Poor	Good	Good	Fair
Training requirement	Low	Low	Low–Moderate	High
Ease of updating information	Very Poor	Good	Good	Moderate–Good

Manual systems.

Manual systems represent the traditional way of managing information. The key to a successful system is proper management of records contained and maintaining easy accessibility to documents and maps. This can be achieved through the establishment of a library system linked to proper document storage and viewing facilities, the latter being particularly important for using maps.

Manual systems have a number of advantages:

- they are familiar to a wide range of people and require no specialist skills to understand and use.

- they enable the user to view a source document in it's original form and context, rather than after standardisation within a computerized system.

- they are perhaps more conducive to a user "gaining a feel" for an area and making subjective judgements than the rigidity of a computerized analysis.

- they may be in an unstructured format, providing no preconditions for use of the data.

They do however suffer from a number of disadvantages:

- they are unwieldy to update and prone to error. The geographic relationships of features in a region as dynamic as the coastal zone are liable to change on a frequent basis. For example, in the case of hydrographic charts, changes are published monthly, which then have to be manually updated on all archive copies of charts. Clearly this can be problematic in terms of map legibility and also managing the logistics of updating all copies of maps held. Care has to be taken that outdated maps are not inadvertently supplied as current material.

- any analysis, particularly that involving the geographic relationships of features is unwieldy, time consuming and prone to error. Consequently, comparisons of information may not be carried out on grounds of time and cost and complex analyses may simply be impractical. The full potential of the information stored is therefore not realised.

- unless properly stored, indexed and managed, access to documents becomes poor. In a number of organisations,

114

commercial pressures on space lead to inadequate library facilities being provided with consequential loss and damage to archive material.

Database development

Simple computerised databases can be developed using readily available personal computer software (eg dBase, Foxpro, Paradox, Delta etc). In their simplest form, databases represent a table of information. Each row or **record** in the table represents an individual feature (for example a borehole, a littoral cell a landslide and so on) and each column or **field** in the table represents a descriptive attribute associated with that feature. Each record should be classified with a unique identifier to enable rapid identification. Table 10.2 illustrates a simple database file structure, each record representing the site of a geotechnical investigation and each field storing salient information about the investigation.

More sophisticated databases may hold information in a series of related database files linked by a common item. This allows the establishment of one-to-many and many-to-one relationships in an efficient manner. For example, the site of a geotechnical investigation may be held in one database file and the results of ground investigations held in a series of related files. The geotechnical investigation site will only occupy one record in it's file, whereas the ground investigation results (eg, piezometer or inclinometer readings from a number of dates) may occupy many records in the associated files. The link is provided by a unique identifier (usually a reference number) identifying and linking all these discrete pieces of information into a coherent and ordered whole. Another common use of related database files is for the production of look-up tables. Look-up tables relate a coded value to it's description. For example a particular rock type can be stored in the main database table as a number, with it's full attribute description stored in the look-up table. The advantages of this method are twofold. Firstly the use of coded values in a main database table promotes efficient storage and also enhances the ability to preform analytical operations on the data set. Secondly, storing descriptive values in a look-up table means that the information only has to be entered (and updated) once for all records of that type.

Assuming proper documentation is made of the database structures properly describing what each database field stores and what each field value means (Table 10.2), a knowledgeable computer user can make direct use of these "raw" database files. However, it is more common practise to develop some form of menu-system whereby non-computer specialists can readily access and use the data. Most proprietary database packages have the facility for a database developer to produce such an interface within a limited time and budget thus maximising the accessibility of the information to the non-specialist computer user.

A fundamental issue in the development of a database is the **standardisation** of information that is to be stored. Here it is useful to develop a **standard proforma** to record **key information** from, for example, different geotechnical investigations, in a systematic manner. Each proforma may contain the following information:

(i)   database reference number;
(ii)   the title and date of the investigation;
(iii)   the grid reference;
(iv)   the organisation that commissioned the work;
(v)   the organisation that carried out the work;
(vi)   whether the data are confidential or not;
(vii)   the nature of the investigation and the type of data that are available;
(viii) brief soil and rock descriptions;
(ix)   an indication of the quality of the data.

Once proformas have been completed, they are entered into the database file structures by keyboard. This is a quick and straightforward task (although repetitive) and can be carried out by junior personnel which keeps down the cost of data capture. Clearly, as with all data capture tasks, a rigorous checking exercise is required to ensure accuracy.

Each proforma will only contain a summary of the original geotechnical investigation. However, if a suitable classification is used, a user will very quickly be able to gauge the relevance and usefulness of the investigation to the task in hand and negotiate with the holder or originator for it's release and use. With regard to the use of third-party originals, a number of issues and trends are worth exploring:

Table 10.2 The structure of a computer database file containing records of geotechnical investigations (from Rendel Geotechnics, 1992).

Structure for database: \STHL\SHHEAD.DBF
Number of database records: 309
Date of last update : 07/11/90

Field	Name	Type	Width	Description
1	GSLREF	Numeric	4	Investigation identification number (as shown on map).
2	REF_SUF	Character	1	Identification number suffix (where required).
3	EASTING	Numeric	5	Grid reference – easting.
4	NORTHING	Numeric	5	Grid reference – northing.
5	TIT1	Character	67	)
6	TIT2	Character	67	) Title of investigation.
7	TIT3	Character	67	)
8	COMBY	Character	67	Organisation commissioning the investigation.
9	CONTRACTOR	Character	67	Contractor.
10	CONSULTANT	Character	67	Consultant.
11	REPREF	Character	20	Report reference.
12	EXEDATE	Character	20	Date of investigation.
13	REPDATE	Character	20	Date of report.
14	DATATYPE	Character	67	Type of data (eg Site Investigation).
15	DATAQUAL	Character	1	Data quality assessment (G=good, M=moderate, P=poor).
16	COLLBY	Character	3	Initials of data collector.
17	DATECOLL	Date	8	Date data collected.
18	MAPREF1	Character	6	1:10,000 series map(s) covering investigation area.
19	MAPREF2	Character	8	Ditto.
20	REPFACT	Character	1	Type of reporting – factual (Y/N).
21	REPINT	Character	1	Type of reporting – interpretive (Y/N).
22	REPDES	Character	1	Type of reporting – design (Y/N).
23	REPOTHER	Character	10	Type of reporting – other (specified).
24	DTSUPER	Character	1	Ground conditions reporting – detailed superficial (Y/N).
25	DTGEOLOGY	Character	1	Ground conditions reporting – detailed geology (Y/N).
26	DTGEOTECH	Character	1	Ground conditions reporting – detailed geotechnical (Y/N).
27	DTOTHER	Character	10	Details in report – other (specified).
28	BR_DEP_MIN	Numeric	6	Depth (m) to bedrock at site – minimum.
29	BR_DEP_MAX	Numeric	6	Depth (m) to bedrock at site – maximum.
30	MAX_BOR_DP	Numeric	6	Maximum borehole depth at site (m).
31	WT_LV_MIN	Numeric	6	Depth (m) to water table at site – minimum.
32	WT_LV_MAX	Numeric	6	Depth (m) to water table at site – maximum.
33	MON_PERIOD	Character	15	Period of monitoring.

- The advent of low–cost scanning hardware and mass media storage devices (eg CD–ROM's) now means that document storage and retrieval now longer presents a significant cost or technological barrier. CD–ROM drives can be purchased for a few hundred pounds and A4 colour 1200 dpi scanners are priced similarly. Coupled to easy to use Windows–based software, the capture, storage, retrieval and transfer of original documentation in computerised format is straightforward. Original documentation can take the form of written documents, maps, photographs and video.

- The increasing use of the **Internet**, part of the much publicised "Information Superhighway", for the exchange of information between users on a worldwide scale will have an increasing influence on the exchange of information in computerised format (Appendix A.5.6). To date more than 20m users are linked to the system via nothing more complicated than a PC, a modem and a telephone line. Costs are low, connection charges of around £25 and usage costs are equivalent to local telephone rates. Users are said to joining the system at a rate of 1m per month. There is already a vast amount of information available through the Internet, much of it well–indexed and accessible to non–specialist users. As the culture grows it is likely that an increasing amount of that information will be of relevance to planners in the coastal zone.

- The major problem with access to third-party information relates to issues of copyright, cost, confidentiality, liability and so on. Unless some wide-ranging agreement can be reached about common access to data sources (which seems unlikely), these issues will remain a significant barrier to information flow.

Geo-referenced databases

The effectiveness of straightforward databases can be enhanced by linking the database software to a computerised mapping package (eg AutoCAD) and storing the information in spatial form. This methodology provides a way of performing spatial analyses and displaying the results in a map format without the need to invest in a sophisticated geographic information system. The major difference between a geo-referenced database and a GIS is reflected in the way that the systems store and analyze information. A GIS stores and analyses spatial (points, lines, areas) and attribute (features, descriptive items) information in an integrated format, whereas a geo-referenced database performs all analyses through the database and uses the mapping features mostly for the display of results. An **appropriately** developed geo-referenced database can provide a very effective spatial analysis tool at a fraction of the cost of a more sophisticated GIS.

As well as the requirement to standardise the classification of information within the database structure, a geo-reference database normally requires that information is compiled on the basis of a **standard unit**, or basic spatial unit (bsu). A standard unit can be considered an area of land upon which planning decisions are made. They can represent administrative areas (eg counties, local authority boundaries), physical areas (eg littoral cells or sub-cells, ecological units), or arbitrary in nature (grid cells). Table 10.3 summarises the advantages and disadvantages of each unit type. The adoption of fixed standard units greatly simplifies the data processing and display tasks. It can also make the system easier to conceptualise by users. Database tables can be designed in which the standard unit represents a single database record. It is important to remember that the standard unit represents the basic building block of the database. Once compiled and entered into the computer, information cannot be analyzed on a sub-unit basis. It is important therefore at the requirements analysis stage (see below) to choose the most appropriate unit.

Data compilation on a standard unit basis can be carried out manually or using a selection of automated techniques. Experience has shown that manual compilation of data, although laborious, is a cost and time-effective option and is sufficiently accurate if procedures for compilation are set out in advance. It is also a non-specialist operation which can be carried out by junior or temporary staff, thus leading to cost savings. Data entry to the computer, being database oriented, is by keyboard.

As part of an **Applied Earth Science Mapping** study of the Torbay area (Geomorphological Services Ltd., 1988) a geo-referenced database was developed storing a range of earth science information on the basis of 50m x 50m (0.25 hectare) grid cells for an area of approximately 75km^2 (7,500 hectares). For each of the 30,000 grid cells within the study area, a range of earth science information was compiled based on the results of the study, namely:

- location – grid reference;
- bedrock geology and structure;
- superficial geology;
- geomorphology;
- slope steepness;
- soils;
- geotechnical conditions;
- sites of geotechnical investigations;
- sites of mineral workings;
- land use planning provisions;
- ground characteristics for planning and development.

The information was stored in a simple database file structure (Table 10.4) as a series of numeric codes. A look-up table linked the coded values with descriptions of the relevant earth science information, providing both abbreviated listings as well as more in-depth background information. A small suite of software accompanies the database tables which allows a user to select a grid reference and instantly have access to all known earth science information at a particular site. This type of information system can be very quickly generated; in the Torbay example, the entire database system was compiled and implemented in less than 2 person-months.

This type of geo-referenced database can be used to generate interpretive and summary maps from the stored information, both quickly and less prone to error than manual overlay techniques. Having a standardised unit for each earth science information "theme" permits rapid identification and mapping

117

Table 10.3 Relative merits of the different types of basic spatial unit.

Physical units	
Advantages	Disadvantages
Boundaries are often permanent and easily identifiable on the ground, from topographic maps and air photos.	Physical units are often not recognised or adopted by agencies and government institutions. Consequently, little secondary data may exist for a physical unit, necessitating a data re-compilation exercise
Issues related to physical, natural resource and environmental planning are commonly related to physical land characteristics.	Physical boundaries need to be delineated from maps, air photos and field surveys.
Physical units can often be considered as discrete ecosystems. It follows that planning decisions based on these units will only have a direct effect on that unit and an indirect effect on "downstream" units.	

Administrative units	
Advantages	Disadvantages
Agencies involved in local planning often use administrative units as the preferred unit.	Administrative boundaries may change through time
As a nationally recognised unit, a great deal of secondary data may have already been compiled on the basis of each unit.	Boundaries that do not relate to physical features are difficult to identify from air photos or in the field.
	Administrative boundaries usually bear little relation to the processes that govern natural resource/environmental management

Arbitrary units	
Advantages	Disadvantages
Arbitrary units are not restricted by administrative or physical criteria, and can be amalgamated if required into either.	In order to provide a sufficient level of detail, arbitrary units need to be small. Consequently a large number are required with consequential overheads in terms of data entry, storage and analysis.

of areas which have particular characteristics (Figure 10.1).

In the Torbay example, a slope map was derived from digitised contours shown on Ordnance Survey 1;25,000 scale maps (Figure 10.2). The slope classes were chosen to highlight local limits for urban development the 11° boundary coincides with the 1:5 limit set by Torbay Borough Council Doornkamp, 1988; Lee at al, 1988).

The Review of Erosion, Deposition and Flooding in Great Britain study (Rendel Geotechnics, 1995a,b,) demonstrates another application of the management of geographic information in a geo-referenced database format. The database structures and associated software (Figure 10.3) are necessarily more complex than the Torbay example reflecting the relationships between recorded incidents of erosion, deposition and flooding and

the need to build in a number of "rules" governing these relationships. Nevertheless, the conceptual model is no more complex, data are collected on the basis of a standardised proforma and the database structures designed around the data model. Being a national overview study it has been possible to locate records on a point (grid reference) basis. This has a number of advantages:

- the spatial relationships between features can be simplified to a single grid reference, rather than the more complex data handling required for linear and areal features.

- the overview scale circumvents the problem of poor location descriptions within source documents (mainly scientific literature and newspapers). A large number of records simply locate records with comments like "flooding in the town centre" or "the river burst it's banks leading to inundation of large tracts of land". Clearly it is not

118

Table 10.4 Structure of Torbay geo-referenced database.

Structure for database: C:\TORBAY\TORBAY.DBF
Number of data records: 29689
Date of last update : 27/10/94

Field	Field Name	Type	Width	Description
1	EASTINGS	Numeric	8	Location of grid cell – easting
2	NORTHINGS	Numeric	8	Location of grid cell – northing
3	M11	Numeric	2	)
4	M12	Numeric	2	)
5	M13	Numeric	2	) Bedrock geology and structural features within grid cell
6	M14	Numeric	2	)
7	M15	Numeric	2	)
8	M16	Numeric	2	)
9	M21	Numeric	2	)
10	M22	Numeric	2	) Superficial geology units within grid cell
11	M23	Numeric	2	)
12	M31	Numeric	2	)
13	M32	Numeric	2	)
14	M33	Numeric	2	) Geomorphological terrain units and features within grid cell.
15	M34	Numeric	2	)
16	M35	Numeric	2	)
17	M51	Numeric	2	)
18	M52	Numeric	2	) Pedological soil units within grid cell.
19	M53	Numeric	2	)
20	M54	Numeric	2	)
21	M61	Numeric	2	)
22	M62	Numeric	2	) Geotechnical ground conditions within grid cell.
23	M63	Numeric	2	)
24	M64	Numeric	2	)
25	M91	Numeric	2	)
26	M92	Numeric	2	)
27	M93	Numeric	2	) Land use planning provisions within grid cell.
28	M94	Numeric	2	)
29	M95	Numeric	2	)
30	M96	Numeric	2	)
31	M101	Numeric	3	)
32	M102	Numeric	3	)
33	M103	Numeric	3	) Ground characteristics for planning and development within grid cell.
34	M104	Numeric	3	)
35	M105	Numeric	3	)
36	M71	Numeric	3	) Reference number of geotechnical investigations carried out in grid cell.
37	M72	Numeric	3	) Details of investigation held in parallel database linked by reference no.
38	M81	Numeric	3	) Current or former mineral workings in grid cell.
39	M82	Numeric	3	)

** Total **   100

Information stored as numeric codes. In-depth explanations of each unit stored in look-up table and text files.

possible to define precise areas using these descriptions, it is simpler and more cost effective to define points of interest on a small-scale overview map.

• the location of place names can be standardised using a digital gazetteer. In this particular example an Automobile Association (AA) gazetteer was used (see Rendel Geotechnics, 1995a).

Using point locators in a geo-referenced database simplifies the data processing requirements and does not require the use of a GIS package to

Figure 10.1 The Torbay geo–referenced information system.

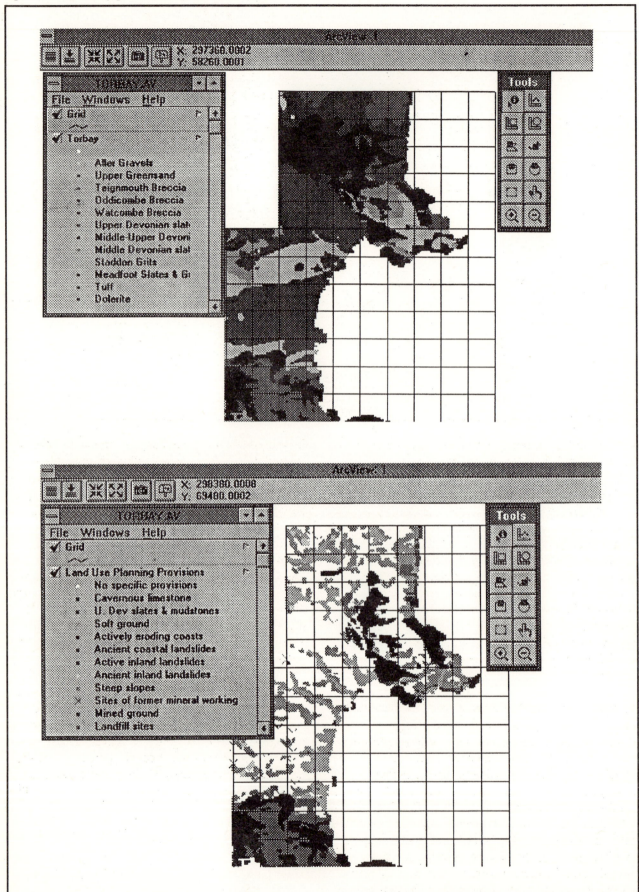

Figure 10.2 Slope map of the Torbay area (after Doornkamp, 1988).

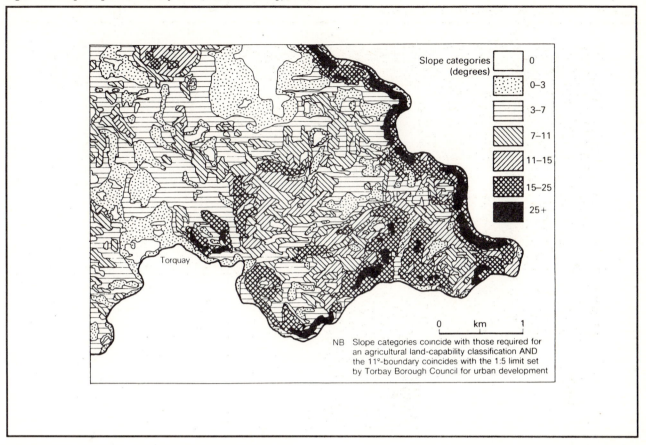

Slope categories (degrees)
- 0
- 0–3
- 3–7
- 7–11
- 11–15
- 15–25
- 25+

Torquay

0 km 1

NB Slope categories coincide with those required for an agricultural land-capability classification AND the 11°-boundary coincides with the 1:5 limit set by Torbay Borough Council for urban development

manage the information. Database information can be transferred to a computerised mapping package such as Autocad. It is important though to consider the problems of the effective map display of multiple records at a single site spread across a range of dates, processes and magnitude. In the example quoted here, a range of cartographic symbols were chosen whereby key indicators were shown by shape (process – erosion, deposition, flooding), colour (geomorphological system – hillslope, fluvial, coastal, aeolian) and size (no. of records at site). Each combination was carefully chosen in order that any symbol/colour/size combination would be cartographically visible at any chosen site.

In addition to transferring map symbols, essential attributes can be down–loaded for storage and display in the mapping package. In this example, each map symbol had a number of attributes attached showing the location, grid reference, system, process, dates, records, events and magnitude of erosion, deposition or flooding. These attributes can be retrieved simply by pointing at the relevant map symbol on the computerised map (Figure 10.4). In the case of multiple records, spread across a variety of dates and events, each record is written to the attributes associated with a particular symbol.

It is important to remember one limitation of this methodology. The down–loading of spatial and descriptive information to the mapping system is effectively a one–way process. Changes made to the database are not automatically written to the mapping package and *vice versa*. In many applications this does not represent a problem, changes can be made periodically and a batch transfer made to the mapping system for display. So long as the database administrator keeps control of updates and makes the appropriate changes to the mapping side of the system as necessary this approach is acceptable. However, a system which monitors continual change to the location and attributes of specific objects and which requires update, for example on a daily basis or even more frequently, and which require the users to have continual on–line access to the latest information is not well–suited to the geo–referenced database approach. A more sophisticated approach involving the utilisation of geographic information system (GIS) software is probably more appropriate in this instance.

121

Figure 10.3 Overview of the Erosion, Deposition and Flooding Database (after Rendel Geotechnics, 1995a,b).

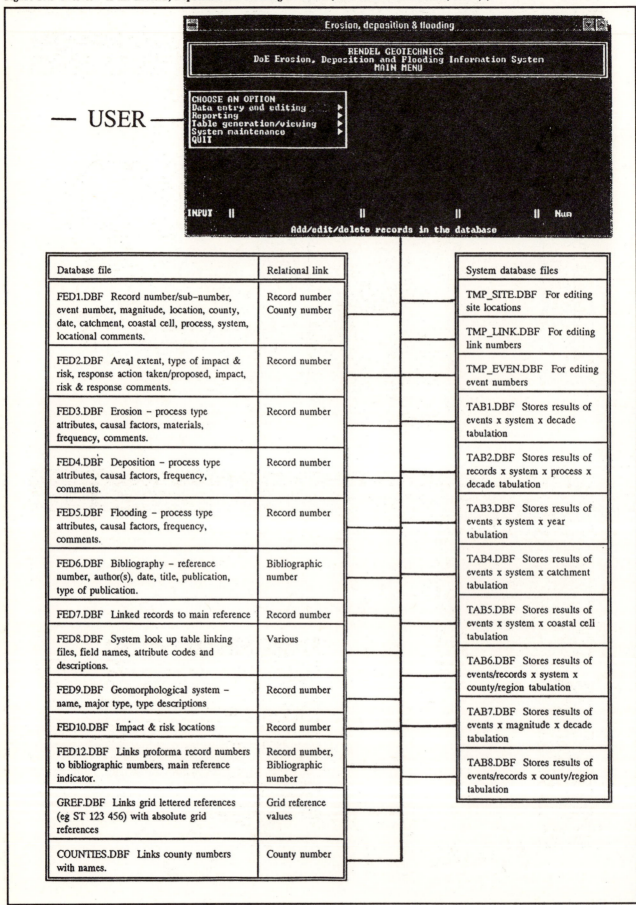

Database file	Relational link
FED1.DBF Record number/sub−number, event number, magnitude, location, county, date, catchment, coastal cell, process, system, locational comments.	Record number, County number
FED2.DBF Areal extent, type of impact & risk, response action taken/proposed, impact, risk & response comments.	Record number
FED3.DBF Erosion − process type attributes, causal factors, materials, frequency, comments.	Record number
FED4.DBF Deposition − process type attributes, causal factors, frequency, comments.	Record number
FED5.DBF Flooding − process type attributes, causal factors, frequency, comments.	Record number
FED6.DBF Bibliography − reference number, author(s), date, title, publication, type of publication.	Bibliographic number
FED7.DBF Linked records to main reference	Record number
FED8.DBF System look up table linking files, field names, attribute codes and descriptions.	Various
FED9.DBF Geomorphological system − name, major type, type descriptions	Record number
FED10.DBF Impact & risk locations	Record number
FED12.DBF Links proforma record numbers to bibliographic numbers, main reference indicator.	Record number, Bibliographic number
GREF.DBF Links grid lettered references (eg ST 123 456) with absolute grid references	Grid reference values
COUNTIES.DBF Links county numbers with names.	County number

System database files
TMP_SITE.DBF For editing site locations
TMP_LINK.DBF For editing link numbers
TMP_EVEN.DBF For editing event numbers
TAB1.DBF Stores results of events x system x decade tabulation
TAB2.DBF Stores results of records x system x process x decade tabulation
TAB3.DBF Stores results of events x system x year tabulation
TAB4.DBF Stores results of events x system x catchment tabulation
TAB5.DBF Stores results of events x system x coastal cell tabulation
TAB6.DBF Stores results of events/records x system x county/region tabulation
TAB7.DBF Stores results of events x magnitude x decade tabulation
TAB8.DBF Stores results of events/records x county/region tabulation

Figure 10.4 Attribute retrieval from Erosion, Deposition and Flooding map (after Rendel Geotechnics, 1995a,b).

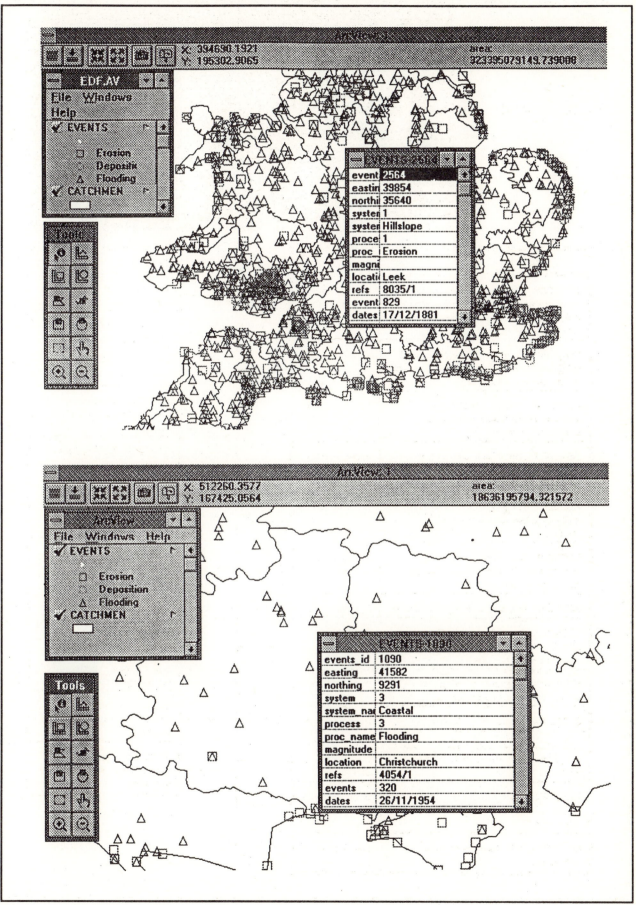

The Potential Use of Geographic Information Systems

Geographic Information Systems (usually abbreviated to GIS) are computer systems which are used to handle spatially referenced data (ie. with a geographical component) with the ability to assist in the storage and analysis of this data and the presentation of the results in map form, (Table 10.5 and Table 10.6).

The review of data relevant to coastal planning and management has found that a large number of spatially referenced data sets are now available in digital or machine readable format (see Appendix A.1). Much information that has traditionally been held as paper maps or charts is currently being converted into digital format. Of note here is collaboration between the Ordnance Survey and the Hydrographic Office on the Coastal Zone Mapping Project (Appendix A.5.1). In the future, survey programmes will both collect and store information primarily in a digital format. For example, the British Geological survey now offer map products in digital form (BGS, 1994).

The current situation would, therefore, seem to be one in which there is a potentially valuable information resource on which a range of GIS applications can be built, and ultimately data may only be available in digital form, rendering GIS even more necessary. Before the potential value of this data resource can be realised, however there are a number of barriers which must be overcome by the user community. The range and character of the data identified by this study suggests that integration if information may be a significant problem. The data sets not only use different conventions and standards, but also have been collected at a range of geographical and conceptual scales. A number of dimensions to the problem of integration can be recognised; these are considered below.

(i) **Technology;** The ability to hold a range of data and to integrate them within a common spatial framework is one of the key features of any GIS. Indeed, it was primarily as a tool for the integration of spatial data that these systems were developed, although until recently significant technological barriers existed which prevented the goal from being fully realised. An assessment of current state of GIS technology suggests, however that the technology itself may **no longer** be the significant barrier to integration that it once was.

The following trends in the GIS market are significant:

- The reduction in costs of hardware, software and GIS services due to the increasingly competitive market environment, coupled with the falling price/performance ratio of hardware and software.

- The growth of the desktop GIS making the technology available on a larger number of platforms.

- The re-engineering of GIS products to conform to widely adopted standards such as MS-DOS/Windows and Unix, and the availability of the GIS product across a range of hardware platforms.

- The development of wide-area networks and client/server architectures which support distributed computer environments that enable integration of systems across geographically dispersed organisations.

- The availability of low cost, large capacity data storage devices (eg. CD-ROMS) so that users are no longer constrained by system resources in terms of the volumes of data they can hold.

- The availability of more user friendly and standardised interfaces to GIS so that the training overhead normally associated with the implementation of such systems is reduced. Users will be able to exploit their skills in particular application areas more rapidly and effectively than has been the case in the past. The technology will be accessible and usable by a larger number of people.

As a result of these factors it is likely that an increasingly large number of users of coastal data will gain access to some form of GIS technology in the short-to medium-term.

(ii) **Data Standards**; despite the significant potential which GIS technology now holds for data integration, the extent to which the potential of the technology can be realised depends on a number of 'external' factors. The first relates to that of the data resources themselves. For data integration to be achieved effectively the user community requires **standardisation** in respect of:

Table 10.5 Typical functions of a comprehensive GIS (after Rhind, 1991).

- Capabilities to retrieve data for any geographical area(s) specified by name or co-ordinates and on any combination of attributes.

- Internal decision rules to assess which operations are feasible given the data available to the system.

- Facilities to transform data from one co-ordinate system to another on an empirical basis using control points and to transform data on the basis of global expressions (eg. from co-ordinates in latitude and longitude to plotting co-ordinates on a specified map projection).

- Facilities to update records for given geographical entities automatically by matching new and old records on the basis of a given, unique identifier and creating file linkages.

- Facilities to convert data in string notation into raster notation and vice versa.

- Facilities to build polygons from supplied line segments. The capability is to be possible where left and right hand attributes are present and where no information other than segment end points is present.

- Facilities to aggregate geographical entities, their geographic representation and their attributes to larger units, on the basis of some built-in hierarchy and on the basis of an externally supplied list, irrespective of the measurement scales of attributes.

- Facilities to disaggregate geographical entities and their attributes, on the basis of the overlay of other entities and specified rules for the allocation of attributes over space.

- Facilities either to carry out simple statistical sampling and summation or to re-format data appropriately for a standard statistical or modelling package.

- Clearly documented interfaces to allow the transfer of data to or from the system at any stage.

- Capabilities to generate tabulated records, graphs, histograms etc. and to map the data at scales and with symbols specified by the user.

- Capability to monitor operations and to build up a composite picture of the user so that default values for parameters are

- data transfer standards
- standard spatial referencing
- standard data quality assurance guidelines definitions and indices

There has been much recent work in the area of data transfer standards both in the UK and Europe. In the UK, the Association for Geographic Information (AGI) is playing a key role, and its National Transfer Format (NTF) for spatially referenced data has recently been published as the British Standard BS7567. In the wider context, developments suggests that the DIGEST format will become the basis of a European Geographic Data Exchange Format. Most GIS vendors now support NTF, so that the task of transferring data held in the internal format of one GIS to that of another **no longer** involves investment in reformatting those data as a 'one-off'.

While much progress has been achieved in the data of transfer standards, difficulties still remain in the development of common spatial referencing systems. In the UK, although the Ordnance Survey National Grid is a widely used spatial referencing system, the datum on which it is based (OSGB36) is not a rigorous scientific one. As a result, positional data collected on a scientific datum such as WGS84 (ie Global Positioning System (GPS)

data) is not integrated with existing map-based information without additional processing. Moveover, the translation between the two is only an approximate one. The OS have yet to publish a standard set of transformations for the UK use community. Within Europe mapping is available on a variety of different map datums and there are significant problems of integration within a common spatial referencing system across international boundaries.

Much earth science data are collected for spatial units such as counties, or districts. Unfortunately there has been little by way of standardisation in respect of these **basic spatial units** (bsu's). Thus, in the development of the DoE's Countryside Information System (see Appendix A.4) by the Directorate of Rural Affairs, for example, the problem has been evident in the extent to which different data suppliers have allocated the 1km x 1km squares of the national grid to counties and countries differently (Nottingham University, 1992). As a result the integration of data from different sources may lead to ambiguity and error in statistical output. **A similar problem exists in respect of different data sets relating to the coastal environment, in that there is no standard for the coastline and no clear understanding of how the representation of the coastline changes**

Table 10.6 The procedures involved in the use of geographic information systems.

The use of such systems involves a chain of activities from the collection of data through to its analysis and use as outlined by the **Chorley Report:**

(i) **Data acquisition and input;** the acquisition of data in both digital and analogue form, map data for processing, digitising and editing and the transformation of these data into a standardised pre–specified format.

(ii) **Data storage;** data formats, hardware configurations, storage media and storage structure and is closely related to the data input and to the data retrieval sub–systems.

(iii) **Data retrieval** is a critical component since it directly affects the user's ability to get at the information to solve a specific problem. Different forms of data are retrievable in different ways. Map data are structured around the cartographic objects (points, lines or areas) often configured into networks or polygons. Non–map data may consist of values for attributes (variables) ordered by entity (observation or record). Database management systems are used to manage both types of data and are usually based on the a data model, the standard models being the based around hierarchical, network, relational or object–oriented concepts.

(iv) **Data manipulation and analysis** operations are frequently embedded into a GIS, although additional capabilities can be added as modules as required. Simple analysis may involve polygon overlay, conversion to different sets of polygons, tabulation of attributes by districts, density and area tabulations and the computation of attribute means. Spatial analyses may consist of terrain analyses, trend surface analyses, spectral analyses, network solutions, location–allocation optimisation, model testing and simulation.

(v) **Output** of information involves similar technology to that of automated cartography. Output may be by means of an interactive display where the user can specify the image on demand or by more traditional hard copy permanent imaging devices. Commonly used devices are screen copiers, electrostatic plotters, matrix printers, pen plotters, laser plotters and inkjet plotters. A GIS may be able to output grid cell maps, network maps, flow maps, base reference maps, isopleth maps and perspective views. In addition statistical output can be displayed in the form of histograms, pie charts and graphs.

(vi) **system management;** addressing the requirements for evaluation, and the organisational aspects of the use and effectiveness of the information transferred through the system. This includes the preservation of accuracy and reliability of data in order to avoid the production of misleading results.

at different map scales.

In addition to the problems which are evident in the area of spatial referencing further difficulties occur in respect of the lack of standards in the area of the quality assurance information published alongside the data themselves. Few guidelines exist, and data suppliers generally provide poor documentation on the issue of accuracy and error associated with their data. Nor is a rigorous comparison of data definitions normally available. The recent **data definitions** project led by DoE for land cover is an exception; this project has cross tabulated the definitions used by all the major recent land cover studies in the UK so that users can more easily make a comparison between them (Nottingham University, 1992).

Although any recommendations regarding the adoption of standards is to be welcomed, it is difficult to imagine how they could be applied, except in their most general form, when each application can be considered unique. Clearly, the priority for an organisation collating information for use at a particular site, at a particular scale and for a particular purpose, will be to ensure that the data collection and classification is appropriate and cost-effective to the task in–hand. There seems

little incentive, particularly in the commercial sector, to tailor the exercise to conform to agreed standards if in doing so the process becomes more complicated, time consuming, expensive and does not directly benefit the ongoing study. Clearly, classifications should included a "minimum data set" for example, location, size, date, and a basic description. Sadly, experience of a number of data surveys of existing sources, for example, the Review of Landsliding in Great Britain (Geomorphological Services Ltd, 1987) and the Review of Erosion, Deposition and Flooding in Great Britain (Rendel Geotechnics, 1995a,b,) show that the standard of reporting, although variable, is often deficient across a wide range of sources. For example, of the 8,835 recorded instances of landsliding drawn from 2,111 reference sources, 60% are unclassified with regard to type of landsliding, 70% are unclassified with regard to age and 95% are unclassified with regard to their activity status. Whilst the inclusion of technical information is not to be expected in non–specialist publications such as newspapers, it is disappointing to see this level of deficiency in scientific publications.

Whilst specific deficiencies in a given data set are generally well-known to the organisation involved in the compilation and are taken into consideration when publishing results, these deficiencies are often less clear to another organisation which wishes to make subsequent use of the data set. Any organisation which wishes to make use of a survey has the responsibility to ensure that the survey is appropriate to their needs. **It is not acceptable to use information simply because it exists**. A validation exercise is necessary to ensure that the secondary source is appropriate. Table 10.7 lists a number of indicators which can be used to assess the suitability of a particular survey.

The lack of clear standards in respect of spatial referencing systems, basic spatial units and data definitions is likely to be a significant barrier to data integration in the short to medium term.

(iii) **Data Availability and User Awareness**; although this study has **identified a wide range of coastal information, the terms and conditions under which data could be made available varies considerably between the different data suppliers.** The market in spatially referenced data is very immature, and few organisations have any experience of it. Examples collected during this project show that very few organisations market data at specific costs (eg: Ordnance Survey and UKDMAP; see Appendix A.2.3), whereas the majority offer data through negotiation. Issues affecting data release include:

- pricing
- security and privacy
- sensitivity
- copyright
- liability
- ownership
- marketing strategy

Projects such as this will help overcome the problem of users being unaware of what data are potentially available. Other projects which will help make 'information about information' (metadata) more generally available include the AGI's database created out of the results of the DoE's Tradable Information Initiative, and the GENIE project. Such work will not, however, result in the more effective use of data unless these other barriers to availability are overcome. **At present the key issues would seem to be those of pricing and the problem of cost–recovery by key agencies such as the OS, and protection of crown copyright.** Both have and are being debated in the GIS community, and there has been a view expressed that the cost of OS data has limited take–up, particularly in the local government sector. The

AGI are currently chairing a round–table which brings together the OS and its customers.

Data costs and leasing arrangements are likely to be a significant factor in affecting the integration of information about the coastal environment.

(iv) **User Needs and Institutional Change**; although the technology for handling environmental data by computers has grown rapidly in recent years, the need to design systems to meet particular user needs remains a significant challenge. On the one hand the data handling requirements of the range of users interested in coastal planning and management issues are not fully understood, although the investigation models developed as the basis of this study may help. On the other, few large scale conceptual models exist to help the policy advisor explore the consequences of policy decisions. The modelling of 3–D processes in GIS is a potential area where development is required.

A review of the possible and potential applications suggests that **no single GIS solution is likely to be appropriate in all circumstances of coastal management.** Users confront problems at a variety of spatial and conceptual scales, and will require different sorts of GIS platform and different levels of baseline information on which to build the data model. Experience in developing the Countryside Information System for the DoE suggests that as central government draws in information from the statutory agencies concerned with the rural environment, for example, the detailed data holdings of those agencies must be processed, aggregated and generalised if they are to be of use to the policy advisor working at the national scales. Their data handling requirements will differ significantly from the data suppliers who require more sophisticated, research–orientated types of GIS.

In order to facilitate the better integration of data across the range of possible users, a much better understanding of the requirements of the different type of users working across the spectrum from the strategic or national scales through to the local is needed. This understanding must be represented in terms of a set of data models which show how the different users conceptualise the world about which planning or management decisions are being made. Moreover, a much better understanding needs to be achieved of the way in which information is used in the decision making process. **The full potential of GIS will not be fully realised if it simply automates existing procedures. Rather, it must**

127

Table 10.7 Checklist for assessing suitability of existing surveys.

Indicator	Comments
Scale	Compiled and measured at a scale similar to that planned for the survey in hand.
Precision	Measured to a level of precision compatible to planned usage.
Comprehensiveness	Includes all key descriptors related to the subject.
Timeliness	Recent survey based on up to date sources.
Accessibility	Usable without restrictions by supplying agency.
Reliability	Of uniform reliability and credibility across the study area.
Credibility	Compiled using a recognised and credible methodology.
Validity	Checked for methodological and data entry errors.
Convenient	In a format appropriate for inclusion with methodology planned for survey in hand.
Relevance	Of sufficient relevance and importance to justify inclusion.

Adapted from: Environment and Development Support Project (EDSP), 1991. Environmental Information Systems for Decision making, Planning and Monitoring. Paper delivered to workshop on "Environmental Considerations in area–based rural development projects". December 1991, Manila, Philippines.

improve the *quality* of decision by changing existing working practices and institutional structures so that more and better information is available for the planner or policy advisor. The development of institutional structures that will allow organisations to fully realise the potential of GIS is likely to be a significant challenge in the future.

An important institutional factor affecting the successful integration of coastal planning data will be that training. The overall level of investment devoted to user education is likely to have an increase, despite the development of 'user friendly' interfaces, because the high level of skill needed to interpret and understand data in the context of the particular applications areas.

The Ventnor GIS

The Ventnor GIS is a pilot computerised system developed specifically as a demonstration of how geographic information systems can benefit the planning process by the integration and presentation of a range of earth science related information in a form both understandable and easily accessed by planners (see Annex B).

Clearly, each coastal site is unique and will have it's own set of priority issues and conflicts between developmental pressures and the natural

environment. In the case of Ventnor, the town is sited on a large, periodically active, pre–existing landslide system that has caused significant damage and associated costs to structures over time within the town. The computerised information system has been developed as a potential aid to local planners in order to integrate earth science related data into the planning system. The Ventnor area has been subject to a number of detailed surveys, including the DoE research contract "Coastal landslip potential assessment; Isle of Wight Undercliff, Ventnor" (Lee and Moore, 1991) and later updates (Rendel Geotechnics, 1995c), consequently a large amount of earth science related information is available for inclusion in the information system.

The system has been developed on a personal computer platform using three suites of software; AutoCAD Release 12, ArcCad Release 1.1 and dBASE IV database software.

AutoCAD and dBASE are amongst the personal computer market leaders in their respective fields, namely that of computer aided mapping and database management. To this end, they are found in a large number of organisations and the data formats produced are compatible with a wide range of other software. ArcCad is a software package providing the GIS link between AutoCAD and dBASE. In GIS terms, this software configuration represents a modest investment – software costs total around £6,500 (1994 price). Additionally once baseline and derived data sets have been established using the above software configuration, they can be viewed and output using a low cost software package (Arcview), a package that is included with ArcCad with further copies costing £1,100 (1994 price).

The minimum hardware configuration for effective use of this software comprises an 80486–DX66 personal computer with 16Mb RAM. Current unit costs of suitably configured PC's are less than £2,000 per unit. Faster Pentium processor based PC's are currently costing around £3,500 per unit (1994 price). In line with current trends prices are expected to continue to fall dramatically.

Peripheral hardware devices such as plotters and printers are also priced very competitively. A0 (1176mm x 840mm) size inkjet plotters are costing around £6,000 (1994 price), A2 (588mm x 420mm) colour printers can now be purchased for less than £1,500 and A4 (297mm x 210mm) colour printers for less than £400.

The Ventnor Information System currently contains information for a 1200m x 800m (96 hectare) area within the town centre. For this area a range of earth science related information has been captured and formatted:

- Ordnance Survey 1;2,500 scale digital map data (see Appendix A.5.2)
- Geomorphological units from a ground survey of the area carried out as part of the DoE Ventnor study.
- Records of structural damage from a ground survey of the area carried out as part of the DoE Ventnor study.
- Records of past landslide events based on a historical survey of newspapers and journals.
- Records of ground movement rates at a number of sites and based on a number of surveys, namely analytical photogrammetry, benchmark surveys, extensometer readings and ground surveying.
- Ground behaviour characteristics for each geomorphological unit derived from the distribution and characteristics of the above themes.
- Planning guidance recommendations for each geomorphological unit derived from the distribution and characteristics of the above themes.

A properly planned and implemented information system can have a number of benefits for the integration on earth science information into the planning process at coastal sites. In the Ventnor case, these benefits can be considered in three broad categories:

- integration of information
- analytical capability
- response to change

(i) **Integration of information**; the integration of all earth science information for any site within the study area – for example, any property within the area can now be assessed for records of structural damage, past landslide events which have affected the area, the recorded movement rates, the geomorphological setting, the likely ground behaviour at the site and the current planning guidance recommendations. This information can be accessed in map form to show the geographic relationships between the site and the earth science information and also in tabular and printed form. If more in–depth explanatory information is required, this can be called up on–screen by accessing the relevant parts of the Ventnor technical report and associated source references (Lee and Moore, 1991).

Another useful form of media, not presently integrated into the system but worthy of consideration, is the inclusion of photographic records at particular sites. This is a particularly useful method of capturing the status of sites and events at particular periods in time before remedial works or demolition mask the effects of previous landslide activity.

Rapid and easy access to this kind of information is seen as beneficial in the processing of planning applications at specific sites, in giving guidance to developers as well as in broader planning terms, such as the assessment of risk across an area. Any number of other related themes of information can be included with the information system to further enhance its capability and usefulness. Examples include:

- the integration of property ownership records, including the scanning and storage of original documentation, plans etc.
- records of remedial and building works carried out on properties through time,
- costs associated with remedial works,
- records of previous planning applications,
- the integration of utilities and facilities information – particularly with regard to those which may have a detrimental effect on the stability of the landslide system if

damaged, namely water supply pipes, drains and sewers.

(ii) **Analytical capability**; an inherent feature of geographic information systems is their ability to quickly process and analyze information across an area. This allows the comparison of a number of types of information across an area and the production of summary or derived themes from a variety of analytical methods. In the Ventnor case, each geomorphological unit has been examined with regard to damage, past events and movement rates, and a ground behaviour and planning guidance category assigned accordingly.

The analytical capabilities of GIS is also useful for identifying and targeting areas. For example, the system can be used to quickly identify areas most suitable for development, to find sites with a contemporary movement rate of more than 10mm per year and so on. The integrated nature of the system permits the identification of areas based on a number of discrete themes.

(iii) **Response to change**; coastal sites are dynamic, the effect of anthropogenic and natural processes on the coastal system often leads to changes in ground behaviour. Additionally, the ground model at a coastal site is often only partially understood. In the Ventnor case, the model of the landslide system portrayed in the information system is based largely on surface information and surveys. It is likely that as more information is gathered, particularly sub-surface information, the definition of the model will change. Furthermore, as new events, records of damage and recorded rates of movement are gathered and input, it is likely that the ground behaviour characteristics and planning guidance classification of individual geomorphological blocks within the system will change to reflect the changing nature of the coastal landslide system. Indeed, major landslide events may change the boundaries and distribution of the current geomorphological units themselves.

Geographic information systems are well-suited to reflecting such changes. New themes of information (such as borehole records or groundwater levels) can be easily added and integrated to the model, changes to the geomorphological boundaries can be rapidly made, analysis of the planning implications based on new information can be quickly run and new maps produced in a fraction of the time and cost of manual methods.

Systems similar to the Ventnor information system need not be particularly expensive, neither do they necessarily require long implementation times. A personal computer based hardware and software specification enabling not only the development of a GIS but also an integrated records and photographic library of associated information is shown in Annex B. Time of implementation will obviously vary from site to site dependent on the requirements of the system, the input and capture of information and the system customisation required. In the Ventnor case, system implementation to date has involved around three man-months of time.

The Ventnor system, as a pilot system, represents a pilot GIS based on what would be considered an "entry-level" system, namely a low-cost GIS implementation on a single personal computer platform. This is appropriate for the site, however more complicated systems, involving extra capital expenditure, staffing and implementation may be more appropriate in other instances. The management problem is to target the right level of system which achieves a sufficient level of capability without excessive cost.

Conclusions and Recommendations

This Chapter has summarised a range of approaches to the effective management of earth science in the coastal zone and it's integration into the planning process. Each style of system will have it's own characteristics and suitability to a particular coastal site and a particular planning and management requirement. When considering how best to manage information it is important to devise an **appropriate** strategy. **Appropriate** systems achieve **objectives** whilst working within the **constraints** of the implementing organisation.

Within an organization, the implementation of an appropriate information system can be summarised as:

(i)   identification of needs,
(ii)   assessment of resources and constraints
(iii)   assessment, classification and validation of component information,
(iv)   selection of appropriate methodology based on (i) to (iii),
(v)   identification, purchase and installation of software & hardware platforms,
(vi)   data capture and validation,

(vii)   customisation of system to customer needs – analysis, reporting, output requirements,

(viii) documentation and training,

(ix)   handover and ongoing support.

Steps (i) to (iii) are designed to achieve a comprehensive analysis of the current situation and to define the wishes of the various parties involved with regard to an improved future situation. This is not a simple linear process and is likely to be subject to a number of reviews and modifications before a final picture emerges. It is often the case that an organisation will only have an outline idea of requirements initially, but it is likely (and hoped) that these ideas will become more focused and realistic as an understanding of the capabilities, limitations and cost implications of particular strategies becomes clearer. Once this stage is reached, it is possible to define a strategy around which system implementation can be based. It is quite possible that changes will be made once a system has been selected and implementation is under way, particularly as the benefits of improved information management potential are realised on a wider scale within an organisation. It is however, important to keep tight control on projects and to remain focused on the prime objectives. Successful project management allows a degree of modification to the strategy without excessive impingement on time and budgetary constraints. It is beneficial to test strategies using a representative pilot area in order that modifications can be made to the implementation strategy without recourse to large–scale changes across the entire study area.

It is not the purpose of this report to make recommendations regarding which implementation methodology is most appropriate in the coastal zone. Each site is unique and must be evaluated on a case by case basis. Rather, this report highlights some of the issues which should be addressed at each stage in the implementation process.

(i) **Identification of needs**; when evaluating needs, a number of issues need to be examined. These could include:

- The area of the site and the scale(s) of operation – is an overview scale sufficient or does the process require detailed analysis at large scales? What is the largest scale of operation? Can all types of information be stored at one scale, or is it necessary to have a variety of levels of detail. What are the ramifications of the area of the site and the scale of operation in terms of quantities of data and the consequences in terms of data collection, capture, storage, analysis and output.

- Is there a basic land unit on which decisions are made? If so, are these units coherent across the study area? If not, is it possible to devise one which satisfies current needs?

- What types of information need to be stored in the system? How are these stored at present? Is there a comprehensive data set in existence? If not, are secondary sources available?

- Current procedures. How is information managed at present? What are the advantages and limitations of the current system? Which areas currently cause bottlenecks in the process? What cannot be achieved at the moment because of constraints and limitations of the system?

- What is required now, and what may be phased in at a later date. How flexible is the data model to cope with subsequent change?

- How should the real world be modelled in the system? What analyses, scientific operations need to be built into the system? Are any highly specialist analyses included which may require specialist software development? Does the data model work in two–dimensions or is it necessary to include the third–dimension? Is it necessary to include temporal capabilities and how can this be best achieved?

- What level of integration is required with data from other departments or organisations? Could costs be shared between departments?

- What accessibility is required to the system? Is a single seat unit adequate, or is a networked multi–seat capability required? How many non–specialist users will need to effectively use the system? How often will they need to access information?

- What security is required for the system? Who will be the users and how many will have authority to change the information stored? How sensitive is the information to be stored? What are the legal implications

- Buck A L 1993. An Inventory of UK estuaries. 7 Volumes. JNCC

(viii) the **Directory of the North Sea Coastal Margin**; the directory is intended to bring together information held within the JNCC, the country nature conservation agencies and other organisations, to provide a comprehensive account of the maritime and maritime interest of the North Sea coastal margin from a UK perspective. Information has been grouped into three broad areas.

- a description of the natural environment;

- an account of the current protected status of coastal and marine habitats, communities and species;

- an indication of activities which have an effect on the North Sea coastal margin.

Each resource is described and some indication is given of its importance and the range of variation around the North Sea coasts:

- Doody J P, Johnston C and Smith B 1993, Directory of the North Sea Coastal Margin. JNCC.

The continuing work on the directory includes the production of a series of regional reports which will provide more detailed resources–based information.

Further details of all the reviews outlined in this section can be obtained from the **Joint Nature Conservation Committee, Monkstone House, City Road, Peterborough PE1 1JY.**

A comprehensive directory of important bird areas qualifying for protection under EC Directive 79/409 Conservation of Wild Birds and other agreements has been compiled by the RSPB. Site accounts are provided for 256 such areas around the UK with descriptions of bird populations and key conservation issues at each site:

- Pritchard D E, Housden S D, Mudge G P, Galbraith C A and Pienkowski M W (eds) 1992. Important Bird Areas in the UK including the Channel Islands and the Isle of Man, RSPB. Sandy. £35.

A.3.3 Land and Water Resources

(i) **Agricultural land classification maps** are based on the extent to which the physical or chemical characteristics impose long–term limitations on agricultural use. The maps, presented at 1:63,360 scale show 5 categories of agricultural land, ranging from Grade 1 (land with very minor or no physical limitations to agricultural use) to Grade 5 (land of little agricultural value). They are obtainable, with an explanatory note from MAFF Publications, London SE99 7TP (Tel. 0181 694 8862) and are £6.25 a sheet.

(ii) **Mineral assessment** maps an accompanying reports for bulk minerals are available from the British Geological Survey. The majority of the studies deal with sand and gravel deposits. Of particular relevance to the coastal zone are mineral assessments commissioned by DoE and the Crown Estate of offshore areas (all undertaken by the BGS):

38	Marine sand and gravel resources off Great Yarmouth and Southwold WB/88/9.
39	Marine aggregate survey: south coast. WB/88/31.
49	Marine sand and gravel resources off the Isle of Wight and Beachy Head. WB/89/41.
50	Marine aggregate survey: east coast. WB/90/17.
51	Marine sand and gravel resources off the Humber. WB/91/1.

Further detail can be obtained from the **Sales Desk, British Geological Survey, Nottingham NG12 5GG.** Items 39 and 50 are only available from Marine Estates, Crown Estates, 78 Pall Mall, London SW1 5ES. Information can also be obtained from the Minerals Division Room C15/13, Department of the Environment, 2 Marsham Street, London SWIP 3EB.

(iii) **Hydrogeological maps** are available from the British Geological Survey (see above) at a variety of scales and displaying a range of information from surface water features, saline water intrusion and aquifer potential:

- 1:625,000 scale maps of England and Wales, and Scotland (£10 each);

- 1:250,000 scale maps of Northern East Anglia, Southern East Anglia and South Wales (£12.50 – £15);

179

- 1:100,000 scale maps of the South Downs, the Chalk of Wessex, Hampshire and the Isle of Wight, East Yorkshire, the Permo–Trias aquifer of S W England, Fife and Kinross, Clwyd and the Cheshire Basin, Eastern Dumfries and Galloway (£10 – £12.50).

(iv) **Groundwater Vulnerability Maps**; the NRA has produced a 1:1,000,000 scale map of groundwater vulnerability in England and Wales, based on available geological and soils information. This map recognises 3 vulnerability classes for major aquifers (high, intermediate, low) and is intended to increase general public awareness of the location of groundwater resources at risk from pollution.

The NRA is currently producing vulnerability maps at 1:100,000 scale in a 3–year programme which will provide greater detail.

Copies of the 1:1,000,000 scale Groundwater Vulnerability map can be obtained from the **National Rivers Authority, Newcastle–upon Tyne X NE85 4ET**, price £5.

(v) The **Irish Sea Study Group** was established in 1985 to examine the environmental health of the Irish Sea. Following an initial survey, specialist groups prepared 4 reports which cover selected topics:

- Part 1 Nature Conservation
- Part 2 Waste Inputs and Pollution
- Part 3 Exploitable Living Resources
- Part 4 Planning Development and Management

Further information can be obtained from the Director, Centre for Marine and Coastal Studies, Liverpool University.

A.4   Land Use

(i) **Land Utilisation Surveys of Britain**; The second Land Utilisation Survey of Britain was initiated in 1960. It produces maps at the scale of 1:25,000 in sheets covering an area of 10 km by 20 km, using the Ordnance Survey 1:25,000 map as the basemap. Sixty–four land–use categories are mapped, divided into the following 13 main groups: Settlement (residential and commercial), Industry, Transport, Derelict Land (including abandoned tips), Open Spaces, Grass, Arable,

Market Gardens, Orchards, Woodlands, Heathland, Moorland and Rough Land, Water and Marsh, and Unvegetated Land. In the Industry group, Extractive Industries and Active Tips are mapped separately. Grass infested with juncus rush, an indication of damp conditions, is designated with a special symbol. Mapped sub–divisions of Heathland, Moorland and Rough Land include Wet Heath, Heather Heath and Dunes.

The field mapping for the Land Utilisation Survey was done at a scale of 1:10,560 and is almost complete for England and Wales. The published maps, an index map, and the Land Use Survey Handbook (Coleman and Maggs, 1965), which explains the principals and practical procedure of the Survey, may be obtained from the **Director, Miss Alice Coleman, King's College, Strand, London, WC2**, or from Edward Stanford Ltd., 12–14 Long Acre, London, WC2 E9LP. For Scotland completed manuscript maps may be inspected at the National Library of Scotland, Edinburgh.

Where the maps of the Second Land Utilisation Survey are not available, reference can be made to those of the first Land Utilisation Survey of Britain carried out in the 1930's under the directorship of L Dudley Stamp. The maps were published at a scale of 1:63,360.

(ii) **Corine Land Cover**; a land cover database has been prepared at an original scale of 1:100,000 in the various European community member states and regions. The database has been compiled from satellite imagery, using 44 classes of land cover arranged in a 3–level hierarchy. The main levels are:

- artificial surfaces;
- agricultural areas;
- forests and semi–natural areas;
- wetlands and water bodies.

Boundaries were digitated and held in an ARC–INFO based GIS system.

Data is available in the form of maps, statistics and ARC–INFO datasets from:

The European Environment Agency Task Force DGX1, Commission of the European Communities, Rue de la loi 200 1049, Brussels, Belgium.

(iii) the **Countryside Information System (CIS)**; CIS has been developed to give policy advisors, planners and researchers easy access to countryside information for each square in a 1 km grid of Great

Britain. The dataset includes the results of the **Countryside Survey 1990** carried out by the Institute of Terrestrial Ecology which studied the current state and recent changes to the countryside, based on field surveys in 1978, 1984 and 1990. The CIS includes a summary of the CORINE Land Cover Map (at 1 km square resolution). Other datasets include:

- vegetation summary data 1990;
- soils summary data 1990;
- geology summary data 1990.

The CIS will run on a MS–DOS system computer with 486 Microprocessor, a minimum of 5Mb RAM and 30 MB free hard disc space.

Further details can be obtained from:

Software Sales
Institute of Hydrology
Maclean Building
Wallingford
Oxfordshire OX10 8BB

(iv) Ancient Monuments;

Lists of ancient monuments scheduled under the Ancient Monuments and Archaeological Areas Act 1979 may be obtained from:

- English Heritage, Fortress House, 23 Saville Row, London WIR 2HD;

- Cadw: Welsh Historic Monuments, Executive Agency, Brunel House, 9th Floor, 2 Fitzalan Road, Cardiff CF2 1UY;

- Historic Scotland, 20 Brandon Street, Edinburgh EH3 5RA.

A.5   Other Sources

A.5.1   Topographic Maps

Ordnance Survey maps can provide a wealth of topographic and landscape information which can benefit coastal planning and management. The most widely used map series include:

(i)   1:50,000 scale Landranger Series of Great Britain, with contours at 10m vertical intervals;

(ii)   1:25,000 scale Pathfinder series of Great Britain, with contours at 10m vertical intervals, often showing the 5m AOD contour;

(iii)   1:10,000 and 1:10,560 scale maps. These are the largest mapping of mountain and moorland areas, and showing contours. The contour interval is 5m;

(iv)   1:2,500 scale, covering most of the country except mountain and moorland areas, and showing contours. Height information is depicted by means of spot heights and bench marks;

(v)   1:1,250 scale, covering urban areas. Height information is depicted by spot heights and bench marks.

The Ordnance Survey and the Hydrographic Office are currently developing a 1:25,000 scale **Coastal Zone Series** of map sheets, presenting a unique combination of topographic and hydrographic information. This series is likely to be of particular value to coastal planners and managers.

Further information on the availability of Ordnance Survey map data can be obtained from **Dept; LM, Ordnance Survey, Romsey Road, Maybush, Southampton SO9 4DH**.

An important feature of Ordnance Survey maps is the availability, through libraries, of older editions of the current map sheets. Comparisons can, therefore, be made of the cumulative changes that have been recorded between map editions; this can be important for determining historical rates of cliff retreat. The first edition of 1:63,360 scale maps (now superseded by the 1:50,000 scale series) will generally date from the period 1805–1840. The survey at 1:10,560 scale began in 1840, with the scale increased to 1:2,500 in 1853, and completed in 1895. The first revision of the 1:10,560 and 1:2,500 scale maps took place between 1891 and 1914. Harley and Phillips (1964) provide a comprehensive guide to these early map editions which may be consulted in local collections (e.g. County Libraries, County Record Offices, etc.); complete sets are held at:

- British Library Reference Division, British Museum, London;

- National Library of Wales, Aberyswyth;

- National Library of Scotland, Edinburgh.

Harley (1972) gives a guide to other types of historical maps. Large-scale tithe, enclosure and estate maps, many of them earlier than those of the Ordnance Survey at a similar scale, are available in manuscript for many areas. From 1836 to 1860 a series of Tithe Survey maps was prepared in connection with the Tithe Commutation Acts. These are very detailed topographic surveys, usually at a scale of 13.3 or 26.7 inches to the mile, and exist for thousands of parishes. One copy of each map was deposited with the Tithe Commissioners, and may be consulted along with the other tithe maps of an earlier date at the **Public Record Office, Chancery Lane, London, WC2**. Other copies may be found in the County Record Offices or County Libraries. The Enclosure Maps, often at a similar scale to the tithe maps, are generally earlier, often dating from the first decades of the nineteenth century. They are best sought at the County Record Offices, where large-scale estate maps may also be found.

A.5.2 Digital Map Data

Digital maps, in computer-readable form are available from the Ordnance Survey. Conventional map details (lines, points, text) is represented as strings of coordinates and is recorded on magnetic media. Once converted to digital format the information is suitable for use in Geographic Information Systems (GIS) and other management systems. The following products are currently available.

(i)   1:625,000 scale, including communication data, settlements, hydrological and coastal features, administrative boundaries and the coastline. The information is available for the whole of Great Britain;

(ii)   1:250,000 scale, offering a greater accuracy than the 1;625,000 scale data and supporting similar applications. It is available for the whole of Great Britain;

(iii)   1:50,000 scale colour data, providing all the details of the Landranger series. Data is supplied in 20 x 20 km National Grid tiles. The product is available for the whole of Great Britain;

(iv)   1:50,000 scale height data produced from the contours on the Landranger series and available as digitised contours or as a

terrain model matrix. It is available for the whole of Great Britain.

(v)   1:10,000 scale black and white data, providing all the details for the 1;10,000 map sheets **except** contours. From 1993 the product has been available for the whole of Great Britain.

(vi)   1:10,000 scale height data, derived from the contours published on the 1:10,000 scale map series. The data is only available to special order.

Further details of the availability and supply terms of the various Ordnance Survey digital map products can be obtained from: **Digital Sales, Ordnance Survey, Romsey Road, Maybush, Southampton S09 4DH.**

A.5.3 Hydrographic Charts

Admiralty Charts of tidal and coastal waters show the topography and composition of the sea bed, the presence of sand banks and mudflats, and the nature of the foreshore. In tidal areas all charted depths are referred to a Chart Datum which is stated on each chart. The charts also show navigation channels and aids which must not be obstructed or obscured, wrecks, submarine cables and pipelines.

Comparisons made between different chart editions can provide an insight into the nature of coastal changes over the last 10 years or so.

Charts are at various scales up to about 1:4,500; further details can be obtained from the **Hydrographic Office, Ministry of Defence, Taunton, Somerset TA1 2DN.**

A.5.4 Aerial Photographs

Aerial photographs are a useful source of information, some of which may not have been recorded on topographic maps. To the experienced eye they can supply much background information about the geology and geomorphology of an area. The full benefit of using aerial photographs can be obtained if they are examined stereoscopically (when an enhanced impression of relief is obtained).

Existing aerial photograph coverage of an area can be obtained from private aerial survey firms (Table A.2), the **Air Photo Cover, Group Ordnance Survey, Romsey Road, Maybush, Southampton S09 4DH** or:

- The Air Photographs Officer, Air Photographs Unit, Department of the Environment, Room 932, Lambeth Bridge House, Albert Embankment, London SE1 7SB;

- The Air Photographs Officer, Welsh Office, Room G–003, Crown Offices Cathays Park Cardiff, CF1 3NQ.

- The Air Photographs Officer, Air Photographs Unit, Scottish Development Department, Room 1/21, New St. Andrew's House, St. James Centre, Edinburgh EH1 3SZ.

The various Air Photographs Units can supply copies of aerial photographs in their collections, including RAF photographs (which commence in about 1945), and Ordnance Survey photographs up to 1969. They also have commissioned aerial photograph cover such as that for the whole of England and Wales in 1969 at a scale of 1:60,000, and that for 'Industrial South Wales' in 1978–9 at 1:25,000. Official enquirers may be able to borrow prints. Prints may be examined in the print library of the Air Photographs Unit of the SDD and the Air Photographs Library of Wales by appointment. The units also operate the Central Registers of Air Photography for photography held by and obtainable from the commercial air photography companies. The Scottish Air Photographs Unit has cover for some 1,600 miles for Scottish coast in colour, using verticals at 10,000 scale and obliques.

Many County Councils and NRA Regions have their whole areas covered by air photography every four or five years, often at a scale of 1:10,000. The County Planning Departments should be able to give details of cover available and where prints may be obtained, and may allow their own collections to be examined.

A.5.5 Historical Records

Table A.3 highlights a range of sources that could provide valuable background information on historical events. The records held by the various operating authorities (e.g. the NRA, coast

protection authorities etc.) will, in many instances, be the most valuable sources of information; it is important, therefore, that these bodies are consulted at an early stage of an investigation. It is important to bear in mind, however, that documents such as journals and diaries can include valuable descriptions of floods and coastal landslide events. Local newspapers are also an important source of information. Their value is enhanced by the fact that they enable a systematic review of erosion, deposition and floods events over long periods. It is important to bear in mind, however, that historical reports, like current ones, tend to concentrate on the events that affect people or property and, hence, most records relate to built up areas. When researching particular events that have been recorded in local newspapers or documents, it is necessary to make a judgement on the reliability of the data source. Potter (1978) suggests that three questions need to be borne in mind:

- what is the nature of the event being recorded, and with what detail, and is it pertinent to the stated objectives?

- who is making the report, in particular what are his qualifications to know of the event i.e. is it a personal observation based on his own experience; an editing of reports from other people, who themselves may have edited the information; a plausible rumour; or a complete invention; or falsification?

- in the light of knowledge of this type of event, is the report credible, in whole, in part, or not at all?

Even a present day report contains edited information. Events discovered from a search of historic sources may on occasions seem to be just as objective, but generally they tend to be more subjective and usually contain much less detail. It should be remembered that most were originally recorded for purposes other than recording the erosion and flooding aspects of the event and its circumstances.

A.5.6 The Internet

Internet, the so-called Information Superhighway, could have enormous potential for coastal planners and managers wishing to access sources of earth science information in Great Britain and Worldwide. The Internet is a network of computer sites which are connected by various means

Table A.2 A selection of commercial air survey companies.

The following companies usually hold the negatives of survey work carried out for their clients and can supply prints at short notice:

(a)   Aerofilms Limited
       Gate Studios
       Station Road
       Borehamwood
       Herts WD6 1EJ
       Telephone: 0181 207 0666; Fax: 0181 207 5433
       (Took over from Hunting Surveys Limited)

(b)   BKS Surveys Limited
       Sales Estimating Department
       Ballycairn Road
       Coleraine
       Co Londonderry
       Northern Ireland BT51 3HZ
       Telephone: 01265 52311; Fax: 01265 57637

(c)   Cartographical Services Limited
       Photographic Librarian
       Landford Manor
       Landford
       Salisbury
       Wiltshire SP5 2EW
       Telephone: 01794 390321; Fax: 01794 390867

(d)   Clyde Surveys Limited (formerly Fairey Surveys Limited)
       Reform Road
       Maidenhead
       Berkshire SL6 8BU
       Telephone: 01628 21371; Fax: 01628 782234

(e)   Meridian Airmaps Limited
       UK negatives now held at:
       The Royal Commission for Historical Monuments
       19 Flemming Way
       Swindon SN1 2NG
       Telephone: 01793 414 100; Fax: 01793 414 101

(f)   J A Storey & Partners (Geonex JASPHOT Limited)
       92–94 Church Road
       Mitcham
       Surrey CR4 3TD
       Telephone: 0181 685 393

including standard telephone lines, dedicated lease lines and even satellite and microwave links.

The most common use of Internet is **electronic mail** (e–mail) with users having their own unique Internet address, (e.g. Surname @ geography. nottingham. ac. uk).

Telnet is a remote login application allowing users to access another computer from their own machine, depending on their levels of access. Common uses include file retrieval using the **file transfer protocol** (ftp) or searching for information using a **Gopher**. Gopher is an application designed to help users navigate through Internet resources; users specify which subject they are seeking information on and the Gopher will search the Internet documents on that subject. When it finds something of interest this can be read, down loaded or sent to an e–mail address.

The World Wide Web (WWW) could become the world's largest source of information over the next decade. It is a hypertext–based information tool allowing the user to explore the web.

Access to the Internet is via a service provider: this involves paying a service company a registration

Table A.3 Historical sources of flood and coastal erosion event information from the 16th Century to present day (from Potter, 1978).

PERIOD	SOURCE	COMMENT
Present day and recent past	NRA Records Coast Protection Authority Records	The NRA, the former Water Authorities and their predecessors would usually produce a report on major flood events. Information may be contained in these reports on: – flood cause and mechanism – estimated discharge – impact If no reports have been prepared, relevant flow and level records may be available. Coast protection authorities are likely to hold records of damaging events, especially in urban and defended areas. In Scotland, RPAs and Regional Councils may have prepared reports on major events.
	Newspapers	National and local newspapers will probably have accounts of significant events, including photographs of damage, etc.
	Rainfall statistics	Information on the climatic events which have caused flooding or coastal erosion can be found in a variety of journals, including: – British Rainfall (1860–present) – Meteorological Magazine (1866–present)
	Surface Water Year Book	Provides a monthly summary of discharges of British rivers.
18th and 19th Centuries (continued over page).	Newspapers	The British Museum Newspaper Collection held at Colindale, London contains all provincial newspapers and London newspapers since 1801. Articles may contain information on: – flood height – extent of inundation – type of cliff failure – damage "Gentleman's Magazine" dates from 1731 and includes information on flood events and notable landslides.
	Local Histories	Local histories are available in public reference libraries or other collections; they may contain information on damaging events.
	Directories	Many counties are the subject of directories with special sections on the larger settlements. Flood and weather phenomena appear in some; for example, a Lincolnshire Directory (White, 1881) contained a list of storm surge heights recorded at Boston between 1791 and 1877. Norton (1950) provides an index of the availability of Directories for particular areas.

fee (around £7.50 – £25) and receiving a connection via a telephone link.

PERIOD	SOURCE	COMMENT
18th and 19th Centuries (cont ...)	Specialist Sources	Examples include: – County Archivist's Offices – Borough records – Town Council records The **Royal Commission on Coast Erosion**, 1906 and the **Royal Commission on Land Drainage**, 1927 contain detailed replies from many local authorities on the problems encountered in their areas.
16th and 17th Centuries	Public Records Office	The vast national collections began to be catalogued in the 19th century. The **Calender of State Domestic** cover the period 1640–1704, and may include details of floods and weather–related events. The **Acts of the Privy Council** provide similar coverage for the period 1542–1631. The nature and extent of the Public Records Office collection is contained in the "Guide to the Contents of the Public Record Office", (1963–1968).
	Diaries and Chronicles	These are a well recognised source and many have been reprinted (see Matthews, 1950 for a coverage of British Diaries 1442 –1942). Famous example include Samuel Pepys (1660–1669) and John Evelyn (1620–1706) for London; Anthony à Wood who provides an almost complete calendar of weather and floods for the 17th century.
	Specialist Sources	Examples include: – Sewer Commissioners and Courts of Sewers records – Ecclesiastical records

Note:

Matthews W, 1950. British Diaries. An annotated bibliography of British Diaries (1442–1942). University of California Press. Berkeley and Los Angeles.

Norton J E, 1950. Guide to the National and Provincial Directories of England and Wales, excluding London, published before 1850. London.

White W, 1881. History, Gazeteer and directory of Lincolnshire. Sheffield.

Appendix B Checklists for Earth Science Considerations

Guidelines for Earth Science Considerations in the Preparation of Development Plans

These guidelines have been prepared to assist those involved in carrying out surveys of an area in support of the preparation of development plans.

1.   **Identification of Vulnerable Settings and Key Issues**

(a)   These issues are commonly associated with particular settings, you should consult with the relevant operating authorities and conservation agencies to establish whether they are significant in the area.

COASTAL ENVIRONMENT	RISKS	SEDIMENT BUDGET	SENSITIVITY
Estuaries	• flooding • channel erosion • sedimentation	• sedimentation in estuary channels may lead to increased navigation problems • disposal of dredgings on land or at sea • sedimentation important to maintain mudflats and saltmarshes as natural flood defences	• flooding and deposition important to maintain and create conservation sites • sea level rise may lead to increased flood risk • channel maintenance and flood defences may affect conservation sites
Coastal Lowlands	• flooding	• sediment supply needed to maintain beaches, sand dunes, mudflats and saltmarshes as natural coastal defences • sediment supply needed to maintain amenity beaches and sand dunes • mineral extraction from beaches and dunes may lead to increased flood risk • sediment may be needed for beach recharge schemes	• sediment supply needed to maintain and create conservation sites • flood defences may affect conservation interests • sea level rise may lead to increased flood risk
Coastal Cliffs	• landsliding • cliff recession	• supply of sediment from eroding cliffs needed to maintain coastal defences elsewhere • supply of sediment from eroding cliffs needed to maintain amenity beaches and sand dunes elsewhere • effects of mineral extraction on coastal slope stability	• erosion needed to maintain and create nature and geological conservation sites • coast protection may affect conservation interests • sea level rise may lead to accelerated erosion rates
Sand Dunes	• wind erosion • flooding	• sediment supply needed to maintain sand dunes as natural coastal defences • sediment supply needed to maintain amenity sand dunes • effects of mineral extraction on erosion and flood risk	• sediment supply needed to maintain conservation sites • wind erosion needed to sustain ecological value • coast protection may affect conservation interests • mineral extraction may affect conservation interests • sea level rise may lead to accelerated erosion

These key issues should be discussed with groups with detailed knowledge of the management of erosion, deposition and flooding in the area:

Risks	Sediment Budget	Sensitivity
• NRA	• NRA	• Conservation agencies
• Local Authority Engineers	• Coast Protection Authority	• NRA
• Coast Protection Authority	• Coastal Defence Group	• Coast Protection Authority
	• Minerals Planning Authority	• Coastal Defence Group